HOGWARTS LEGACY

COMPLETE GUIDE

BEST TIPS, TRICKS, WALKTHROUGHS AND STRATEGIES

I0479225

Emerald Murazik

Welcome to the ultimate guidebook for the magical world of Hogwarts Legacy! This comprehensive guide will take you on a journey through the wizarding world, providing you with all the tips, tricks, and strategies you need to master this incredible game.

With detailed walkthroughs, expert advice, and stunning artwork, this guidebook is the perfect companion for any Hogwarts Legacy player. Learn how to navigate the sprawling grounds of Hogwarts, master powerful spells and potions, and uncover hidden secrets throughout the wizarding world.

Whether you're a hardcore Harry Potter fan or a casual gamer, this guidebook is packed with everything you need to become a legendary witch or wizard at Hogwarts School of Witchcraft and Wizardry. So why wait? Pick up your wand and join us on this magical adventure today!

TIPS AND TRICKS

With so much to explore and unlock across the wizarding world of Hogwarts Legacy, it makes sense to come equipped with an arsenal of handy tips and tricks to ensure you get off on the right foot with your fellow housemates.

So, whether it's solutions for managing your inventory, tips for causing havoc in The Highlands, or even unlocking cosmetic transmogs, our comprehensive tips and tricks below will have you prepped for everything Hogwarts Legacy will toss your way.

✧ Essential Tips and Tricks for Hogwarts Legacy

As hard as it may be to resist the urge to explore early in the story, the game, unfortunately, prevents you from leaving the quest area, as any exploration outside these bounds will result in your death.

Don't lose valuable resources. Unfortunately, loot isn't automatically collected, so remember to pick up any resources enemies may drop.

If you need to level up quickly to wear gear or complete a particular quest, start searching for Field Guide Pages. It's a fast and easy way to level up quickly.

Gear stats increase as your character level increases, so frequently check your worn gear to ensure you're wearing the best gear possible.

Seek out Eye Chests to earn some quick money, as these mysterious chests will reward you with a whooping 500 Galleons each.

Inventory management will play a significant part in your Hogwarts Legacy experience, as right from the start, you're provided with just 20 gear slots. While it may seem like a lot, you'll soon learn that loot chests are everywhere, and with gear stats increasing as you level up and progress through the game, you'll constantly be destroying and selling off gear pieces. To tackle this issue, you'll want to unlock additional inventory slots by completing Merlin Trials, which are unlocked after completing a mandatory main quest.

Once the Merlin Trials quest is completed, these unique puzzles will appear on the map, each providing a unique twist to complete them, whether with a new spell or by using the environment to your advantage. Each completed trial will count towards the "Complete Merlin Trials" Challenge under the Exploration menu Every milestone completed will reward you with four additional inventory slots – complete all challenges, and you'll have a total of 40 gear slots available.

Unfortunately, Merlin Trials are locked behind a mandatory main quest that is at least a fair few hours into the game if you're mainlining the story, so keep that in mind.

Running out of inventory space? Instead of destroying gear, be sure to frequently visit vendors and offload any unwanted items to earn Galleons. We highly recommend stocking up on Galleons early, as potions and resources can quickly become quite costly.

There's fall damage, so be extra cautious when traversing steep areas.

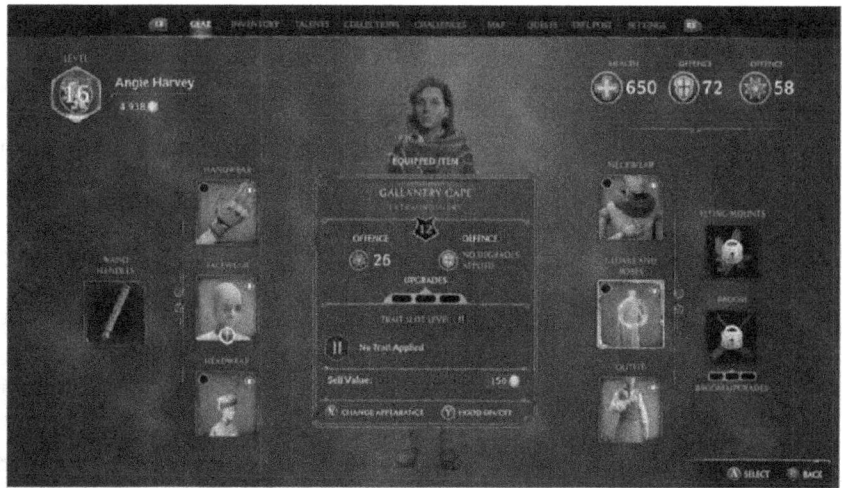

Many of the rewards you receive for completing Quests and Challenges aren't actually gear but instead just appearances. The reason for this being is that Hogwarts Legacy has a transmog system right from the start, but it's kind of hidden within the Gear Menu. So, to enable transmogs, you must have a piece of gear for that slot equipped, like gloves; then, on the main gear screen, hover over the item slot, and you'll now find an option to Change Appearance. Using the transmog system will allow you to maintain your preferred appearance without losing out on vital gear stats that will carry you through combat encounters.

Do note, though, changing out the item will default the look to its original form, so if you want to keep a certain look for your character going, you'll need to transform it every time you switch something out.

As you progress through the main story, your connection to Ancient Magic will allow you to unlock Talents, which are unique enhancements for your combat powers, core abilities, stealth, spells, and much more. Talents are unlocked with Talent Points, with one Talent Point being rewarded each time you increase your Wizarding Level, starting at level 5. Luckily, if you are over-leveled, you will be compensated Talent Points for each level over 5.

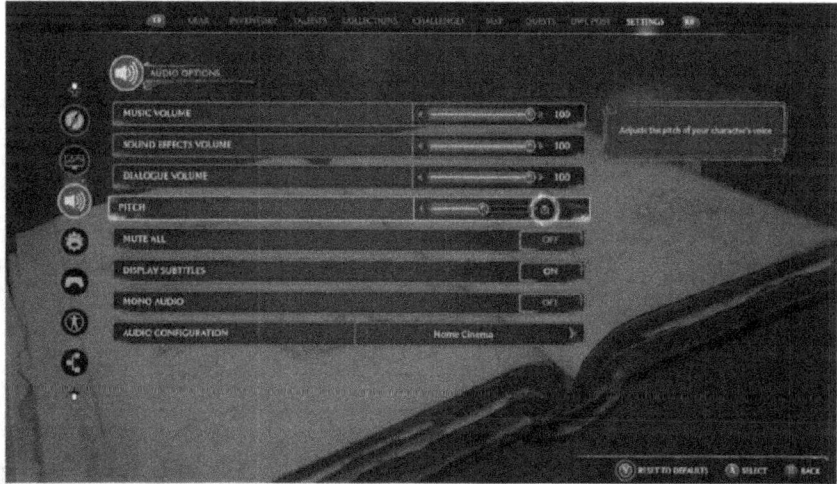

If you change the pitch of your voice in the character creator, you may find that it sounds a bit like a robot since a post-processing filter was applied. Thankfully, if this sound annoys you, we recommend keeping your voice at the default pitch. If you wish to change your character's pitch after the initial setup, you can adjust it under the Audio Options when visiting the Settings menu.

Every character you create has their own save slots. If more than one person is playing on the same account, or if you want to make a character of a different house, you can keep those files separate.

Frequently check the Challenge menu for any completed Challenges, as it's easy for rewards to go unclaimed. Without claiming your rewards, any gear transmogs or upgrades, such as inventory space, won't automatically be available for you to access.

When available, complete all Professor Assignment Side Quests. These essential side quests will be crucial to unlocking new spells.

When tracking a main quest, its bright gold-colored waypoint will always take priority over your own personal waypoints on the mini-map – these are colored purple. These personal waypoints won't be displayed until you have unselected the main quest.

With that said, when viewing the Quest Menu, you'll be able to preview all available quests you can complete. These quest descriptions will provide all the requirements needed to complete the quest and the reward you receive. We highly recommend that you use this information to prioritize which quests you wish to complete first.

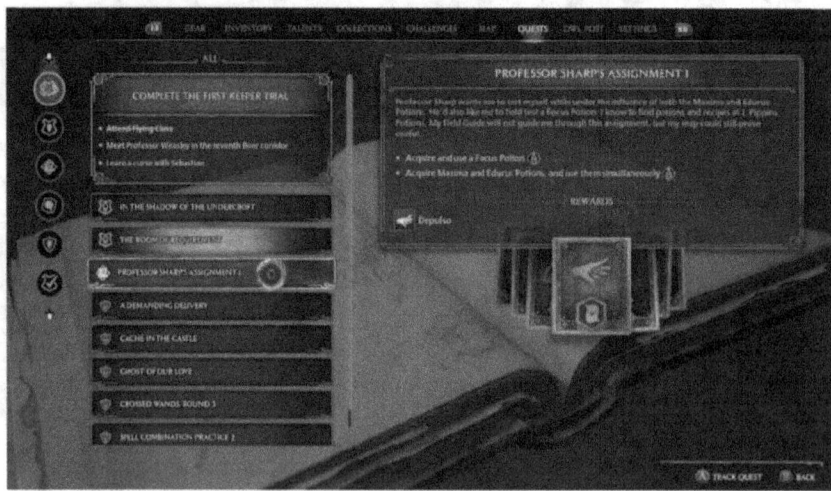

Use the "Locate on Map" button after picking a quest you want to do. While this won't work for Assignments, most other things will be auto-scrolled to on your map, and then you just find the closest waypoint to fast travel to.

Once you've unlocked the Alohamora – the spell needed to picklocks – keep on the lookout for Disguise Moons – which are small monkey statues holding glowing moons. These statues can only be collected at night and are required to increase your lockpicking level.

Want to be mysterious? You'll find an option to toggle your Hood on and off when hovering over the Cloaks and Robes equipment slot.

✧ Essential Combat Tips and Tricks

Rather than firing off spells individually, focus on combos. For example, Incendio can only be done at close range, so you can use Levioso to start a juggle with normal spell shots, use Accio to yank an enemy towards you, and then use Incendio to set them alight.

Other than Ancient Magic attacks or Stupefy, enemies' shields can only be broken using a spell of the corresponding color: violet, yellow, or red. This can easily get overwhelming, so it may be a good idea to match each face button to a specific color, even across different spell sets you can eventually switch to with the D-pad. For instance, on Xbox, you can use B for red spells, Y for yellow, and X for violet, since it's close to blue. Just find a system that works for you and stay consistent!

Speaking of ancient magic, the meter fills up quickly while comboing with your basic shots, so make sure you use it. Throwing objects at enemies (especially the red barrels) or pressing both bumpers at the same time can break shields and do massive damage, often even killing an enemy outright. You can even spec into some Talents that fill the meter every time you perform certain actions.

Deal with enemies standing above you first. They love to stay at the edges of the field and pelt you with stuff, so bring them down to your level quickly!

Utilize stealth when you can! Using the Disillusionment spell and Petrificus Totalus, you can sneak around enemy encampments and stealth takedown enemies, eliminating them from any future fights. You can also use Invisible potions during fights to activate stealth even after being seen.

Visit the Enemies Collection menu to discover important information about certain enemies, including techniques to utilize, weaknesses to exploit, and certain spells that an enemy may be most vulnerable to. Some of these are very context-specific, such as blocking a Troll's boulder to send it right back at him.

It's easy to only focus on the symbols over your head, so you know whether to counter or dodge. But enemies can use AOEs as well, even jumping high into the air or burrowing underground to disguise their attacks. So keep your head on a swivel, and keep an eye on a good escape route if things get too hairy.

Almost every time you parry, you should be holding the counter button down to ensure you cast Stupefy. You can improve Stupefy with certain talents, such as giving it the ability to curse enemies, or even due damage on its own, outright.

You can't interrupt a spell casting motion once you start, so make sure there's not an incoming attack waiting for you before you cast. Hitting an aggressor with Stupefy or lifting groups of enemies can ensure you have time to get spells out.

That being said, keep in mind that Stupefy will launch at whomever you're targeting, not necessarily who you just blocked, so you can use attacks from enemies on the periphery to activate a Stupefy for your most troublesome target.

Stupefy pairs perfectly with the Stunning Curse, Stupefy Mastery, and Stupefy Expertise Talents, so if you're a defensive caster, you may want to prioritize these Talents.

If you care about your appearance, keep an eye on the Dueling Feats at the bottom right of the screen. Fulfilling these requirements will get you closer to unlocking new cosmetic options, and they're just fun ways to try new things in combat.

Equip the item you think you'll need before starting an engagement, if possible. Of course, you can always open the item wheel mid-combat, but going in with a solid plan at the start, whether it be a potion or plant, can save your fumbling fingers from wasting valuable resources. Those plants don't grow on trees, you know!

Talents are extremely important in tuning how you play, but they won't unlock until you almost complete the main story quest Jackdaw's Rest.

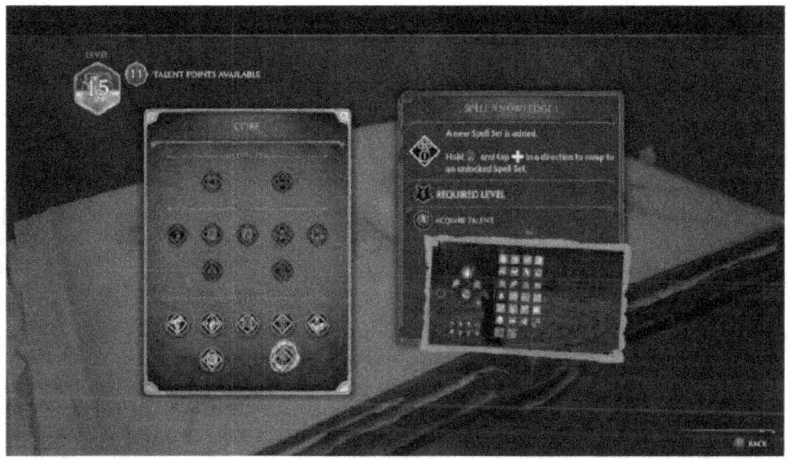

Don't stress too much about managing your spells, though, as you can unlock further Spell Sets with Talent Points once you've completed the Jackdaw's Rest main quest – this will allow you access to four brand-new spells at the press of a button.

While in combat, keep an eye on the bottom right of the UI for Dueling Feats, such as using particular spells or items during battle. Completing these will unlock new cosmetics faster.

Keep in mind that Stupefy will launch at whomever you're targeting, not necessarily who you just blocked. You can use one enemy's attack to stun a more troublesome enemy if you so choose.

✧ Essential Exploration Tips and Tricks

Remember to frequently utilize the Revelio spell, as it's essentially a free X-ray vision. This spell can help detect nearby loot (gold), enemies (red), and quest-related items or puzzles (blue); it'll even help show the solution to hedge mazes.

Throughout the world, you'll discover small loot chests that are often easily found. But don't sleep on them, as while most will simply contain Galleons; there is a rare chance they may contain gear and cosmetic transmogs.

Hogwarts Legacy is full of secrets, so be sure to read any Field Guide Pages you discover. Some will even drop clues to use certain spells on statues to uncover secret passageways from Hogwarts to Honeydukes.

Travel off the beaten path, especially within caves and dungeons, to discover unmarked loot chests - these can sometimes contain valuable items you can later sell to vendors for quick and easy money.

Looking for Demiguise Statues? When exploring Hogwarts, check Classrooms and Professor rooms – such as Professor Fig's – as there's a good chance you will find some of these statues hiding in these areas.

Almost nowhere is off limits in Hogwarts Legacy, as the wizarding world is hugely explorable. So, if you notice a door, be sure to check it out, as chances are you'll likely be able to explore inside.

Make sure you always have inventory space when searching chests. Should you search a chest with a full inventory, there's a chance you won't be able to claim the reward, as the chest will seal shut.

THINGS TO DO FIRST IN HOGWARTS LEGACY

With the world of magic at your fingertips, it can be difficult to resist the urge to explore everything Hogwarts Legacy offers early on in the game. To make your transition as a new Hogwarts student easier, we've compiled a list of essential things to do before you go around exploring.

Acquiring your wand, unlocking essential spells, and buying your first broom are things that will make exploring the world of Hogwarts Legacy more effortless. While you can explore the castle and the rest of the Scottish Highlands almost immediately, you may benefit from seeking out a few points of interest first.

✧ Attend Your First Several Classes

Resist the impulse to explore right away and attend your first several classes—completing them and some early main missions will unlock basic spells, side quests, and valuable items that'll help you scour Hogwarts, Hogsmeade, and the Highlands in greater detail.

Spend Time Learning Spell Combos

Different spells require a secondary attack to make them worthwhile. For example, Incendio can only be done at close range, so combining it with Accio to pull the enemy closer will make the execution much easier

You can practice learning multiple spell combos by speaking to Lucan Brattleby by the Clock Tower Courtyard Floo Flame.

✧ Unlock Floo Flame Locations For Easy Fast Travel

Exploring Hogwarts and admiring the magical interior is a massive part of the fun, but going up and down the Grand Staircase to go from one room to another can get tiring quickly.

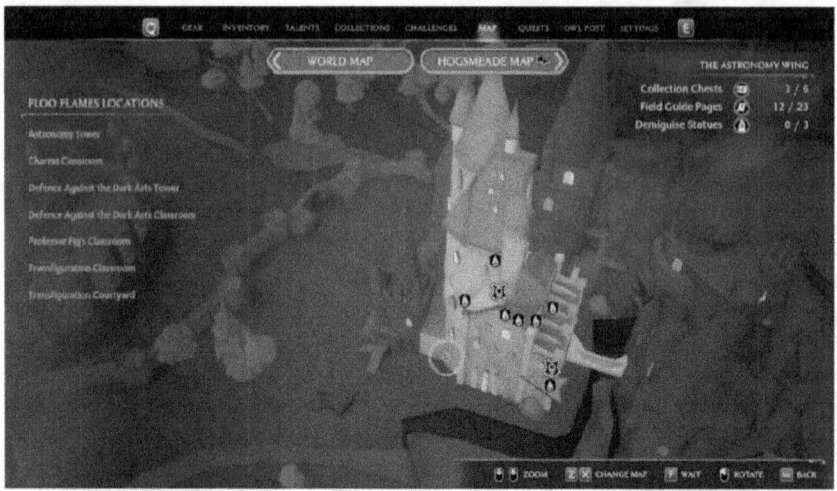

Accessing many Floo Flame locations early in the game will make handy work of traveling from one side of the castle to another, allowing you to track activities and challenges much quicker.

These fast travel points are available everywhere in the game's world, so check out your map as often as possible.

✧ Gain Early XP from Field Guide Pages

After unlocking spells like Accio, Levioso, and Reparo, you'll be able to open up new and exciting quests in and out of Hogwarts. In addition, with those spells, you'll able to acquire Field Guide Pages outside of the ones you can only find with Revelio.

Field Guide Pages give you +80 XP, which is substantial in the early levels.

✧ Craft Your Wand

What's a wizard without their unique wands? Having your own wand will allow you to deal more significant damage and equip it with the Wand Handles you collect during your journey. As soon as you get to Hogsmeade for the first time, you'll want to head to Ollivanders southwest of The Three Broomsticks.

✧ Interact With Everything

The beauty of a magic castle is that almost everything is imbued with charms and spells, which means there is a good chance you'll uncover secrets, collectibles, and equipment just by interacting with random items you see while exploring. For example, interact with the owl statue by the eastern corridor to reveal a hidden room with a treasure chest; or the book inside Tombs and Scrolls in Hogsmeade to reveal a hidden staircase.

✦ Complete the Merlin Trials

You'll start the game with an infuriatingly small number of slots in your inventory. Completing these trials increases your inventory size; otherwise, you will need to destroy or sell some items to make space for new ones.

You can complete your first Merlin Trial during the Trials for Merlin main quest in Lower Hogsfield—you'll need to finish two of the trials to gain your first gear bonus slots.

✦ Change Your Appearance

Calliope Snelling: Oh, come in, come in. Welcome to Madam Snelling's Tress Emporium!

There's a good chance you might not be thrilled with your character's appearance. Don't worry; Madam Snelling's Tress Emporium in Hogsmeade can give you more than a fresh cut—you'll be able to change even your complexion and eye color inside his shop.

✧ Sell Your Unwanted Gear

Without a proper dismantling mechanic for your gear, your best bet at getting rid of old equipment is simply selling them. You can do this at any shopkeeper by clicking on the treasure chest symbol after checking out their inventory. Not only will this make more space for new gear, but selling them will be your fastest way of getting gold early in the game.

✧ Buy A Broom

Though you are free to explore the world of Hogwarts Legacy as soon as you get the opportunity to leave the castle, it's best to mainline the story until you get a chance to buy a broom of your own.

Once you've attended your first Broom Class with Professor Kagawa, head to Hogsmeade's Spintwitches Sporting Needs and speak to shopowner Albie Weekes—all the brooms available will cost you 600 Gold each; not a bad price for the ability to fly around the map.

WALKTHROUGH

✧ The Path to Hogwarts

The Path to Hogwarts is the first main quest within Hogwarts Legacy; where your journey begins, with an ambush on your way to Hogwarts revealing a sinister scheme, and innate ancient powers within you that anyone could have expected.

This guide acts as a comprehensive breakdown of The Path to Hogwarts quest within Hogwarts Legacy, including a full walkthrough and coverage of any optional objectives.

The Journey Begins

After creating your character and pressing "Start Your Journey", a cutscene begins where your player character is speaking with Professor Fig as you prepare to depart for Hogwarts. A newly Apparated George Osric from the Ministry of Magic arrives in the nick of time, and joins you on the carriage ride.

Whilst in flight, George reveals a newspaper discussing the Goblin Rebellion led by Ranrok, and George reveals that Professor Fig's wife, Miriam, warned him about how dangerous Ranrok truly is before she passed away, whilst also sending him a strange container.

Your player character notices a glow on the object the others don't, and upon touching it, it opens to reveal a portkey. Before anyone can truly react, a dragon appears and rips the carriage in two, killing George.

To avoid being burnt by the dragon's fire, you and Professor Fig jump from the carriage and plummet down, grabbing the key and apparating away before the dragon can catch you.

Follow Professor Fig

Landing in a cave, a tutorial begins introducing you to basic movement and navigation functions, as well as the healing mechanic of the Wiggenweld Potion. As you and Professor Fig exit the cave, you realize you are on the coastline somewhere in the Scottish Highlands, with a set of ruins in the distance.

Navigate around the cliffside, following Professor Fig as he provides context to the myth of ancient magic being wielded by a powerful few. Climb the small ridges as you get to them, before you arrive at a wall of enchanted stone. Fig will instruct you to use your Basic Cast a number of times to shatter the wall. Cast it 3 times to break the wall, allowing you to progress.

Slide down the ramp ahead and continue to follow Fig. As you vault up and double back on yourself, you can find the first small chest to loot tucked around the corner to the left (instead of following Fig to the right) which will contain a small amount of Gold.

Follow Fig to the dead end and he will cast Reparo to rebuild the crossing, allowing you to reach the ruins.

Explore the Ruins

Once inside the ruins, investigate the mural ahead of you, and then look left to see a statue. After investigating that, look to the right and loop around the outside of the building to arrive at another section of enchanted stone.

Approach it to see a room on the other side of the wall, which Fig cannot see. After a short conversation, interact with the symbol on the wall to teleport both yourself and Fig to the room you could see through the stone.

Wake the Goblin

Now teleported to this spacious room (which is revealed to be a private section of Gringotts), approach the desk directly ahead of you to interact and awaken the goblin. He will be surprised to see you, asking for vault number 12, and the key to it - Miriam's portkey.

You climb aboard the mine cart and descend deep into Gringotts, with vault 12 being one of the oldest vaults in existence. After passing a dubious-looking Goblin Guard, you finally arrive at vault 12.

Follow Professor Fig/Learn Revelio

Head straight forward for a short cutscene of the vault door being unlocked. Before passing through, you can loot a small chest on the right-hand side. Once inside the vault, the Goblin will seal it behind you. To find a way forward, Fig teaches you Revelio.

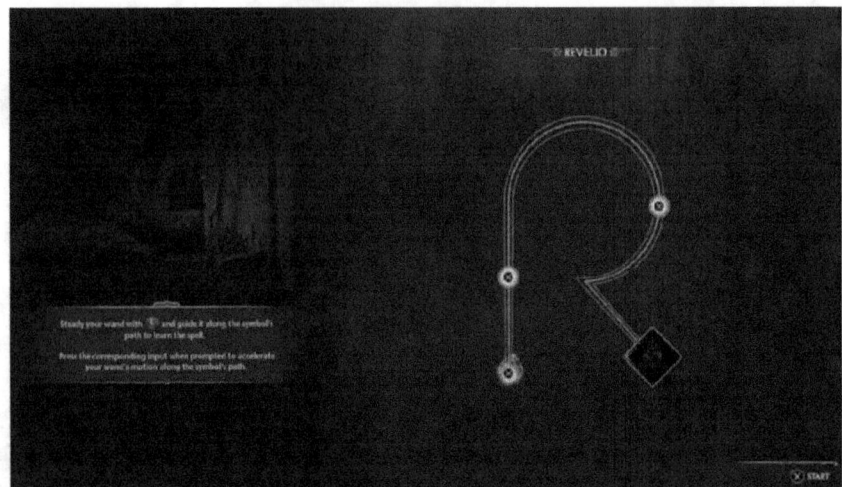

In this mini-game, simply guide the cursor along the tracklines laid out (in the shape of an R), hitting the button prompts as you arrive at them. Keep moving to evade the red shade following your blue magic all the way to the end, and you can then cast Revelio. Get a little closer to the opposite end of the corridor and cast Revelio again to reveal a hidden door.

Proceed Into the Vault

Investigate the hidden door to teleport once again, this time to a cavernous dark room. Stay close to Fig as he casts Lumos, and you'll arrive at another glowing symbol, this time on the floor.

Determine How to Proceed/Learn Lumos

As you interact with the symbol on the floor, the ground will change to the enchanted stone-texture, with a reflection of a statue on the underside. Cast Revelio near the reflection to make the real statue appear. With the reflection following the light of Lumos, Fig decides you should learn it to complete this puzzle. Repeat the Revelio mini-game, this time in the shape of an inverted V, and you can then cast Lumos.

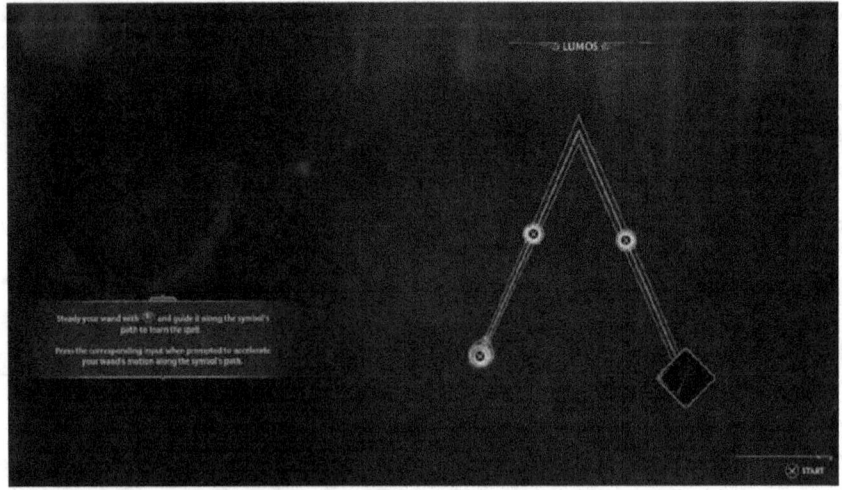

Solve the Puzzle

Cast Lumos and then guide the reflection of the statue to line up with the real one, causing it to stand up and summon multiple other statues. As the statue swings at you, hit the prompt to cast Protego.

You'll then need to cast Protego to deflect 3 incoming enemy attacks, signified by an orange halo around your head. After doing so, use your Basic Cast to destroy the rest of the statues, before Fig disappears and you are plunged into darkness. Cast Lumos and follow the wisps of light to another symbol on the ground.

Solve the Second Statue Puzzle

After activating the ground symbol, cast Revelio to reveal a further 3 statues and their reflections. Use Lumos and find the sweet spot on the ground which causes all 3 statues to align with their reflections at the same time.

Having successfully done so, you'll need to defeat them all using Basic Cast, and protecting yourself with Protego. You can also hold down Protego to trigger a Stupefy counter-attack on a successful block, which stuns enemies, causing them to take additional damage.

Finding Professor Fig

You'll need to trigger the Stupefy counter-attack 3 times and then defeat the remaining statues to progress. Again follow the wisps using Lumos to arrive at a large statue of the same symbol, and interact with the magic on the ground.

Doing so will reveal an archway portal, which you need to proceed through and approach the Pensieve in the center of the room. Interact with it to retrieve the floating container and reunite with Fig. You both view a memory in the Pensieve where Percival Reckham and Charles Rookwood discuss an ancient power that a witch or wizard will eventually earn and master.

As you exit the memory, Ranrok and his troops arrive in the vault, demanding what you've learned, but Fig rejects the proposal, getting into a magic battle with Ranrok that causes the vault's defenses to activate. A giant statue emerges from the ground, battling with Ranrok and the other Goblins, as you and Fig use another gateway portal to arrive in the forest not far from Hogwarts. After agreeing this is where you are meant to be, you both depart for the Sorting Ceremony.

✧ Welcome to Hogwarts Walkthrough

You've finally arrived at Hogwarts as a fifth year, so now it's time to be sorted into a house. The Sorting Hat will ask you a few different questions, and your answers will form a suggested house. However, you're still free to pick from Gryffindor, Slytherin, Hufflepuff, and Ravenclaw at the end. For more information, check out our Best House: What House Should You Choose? guide. Once you've made your decision, a teacher will escort you to the Common Room of whichever house you pick. For this guide, we'll be playing as a Gryffindor.

The next morning, head down to the Common Room and introduce yourself to some other students. In the case of Gryffindor, it's Cressida, Garreth, and Nellie you're looking for. Simply look for the three objective markers within the Common Room and strike up a conversation with them all. Once you're done chatting, leave the Common Room via the main entrance and meet Professor Weasley. In the next cutscene, you'll be given the Wizard's Field Guide and told to follow Professor Weasley.

On the floor below, you'll approach a painting and be ordered to cast Revelio on it to collect a Field Guide Page. Continue to follow Professor Weasley until you unlock a Floo Flame fast travel point. You'll then be told to open up your map and go to the Central Hall.

Go down the steps to the fountain, then have a conversation with both Professor Weasley and Professor Fig. The game opens up from here, allowing you to follow two different Main Quests in Charms Class and Defence Against the Dark Arts Class. You can continue the core story through either of these missions or choose to explore Hogwarts.

✧ Charms Class Walkthrough

Charms Class is the third main quest within Hogwarts Legacy; where you begin your training at Hogwarts by learning the Summoning Spell, Accio.

This guide acts as a comprehensive breakdown of the Charms Class quest within Hogwarts Legacy, including a full walkthrough and coverage of any optional objectives.

Beginning the Lesson

After tracking Charms Class, head to the Charms classroom. Once you enter and begin the lesson, you'll meet Natsai Onai, as well as Professor Ronen. Here you will learn Accio, in the shape of a semicircle, so complete the spell-learning minigame to successfully cast Accio.

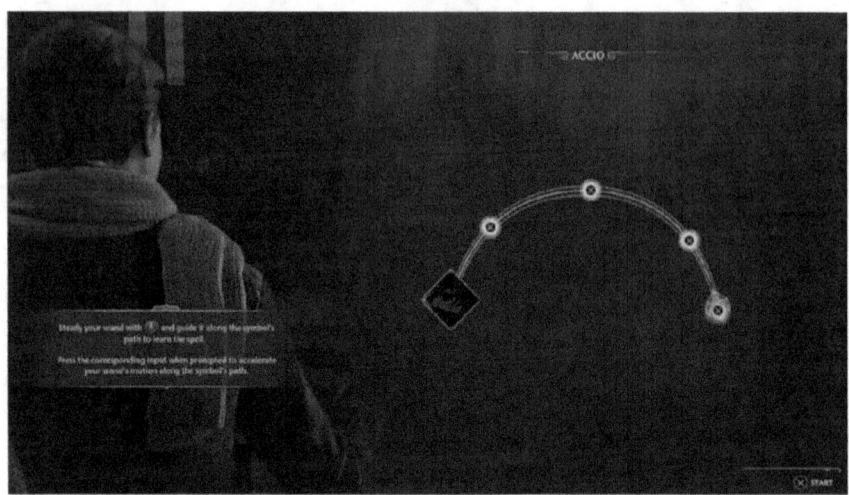

Summoner's Court

Ronen will then escort the class outside, where you will participate in the Summoners Court minigame for the first time. Aim to beat Natsai if you can, but it doesn't matter either way. A second round of the minigame will then begin, now with moving obstacles.

Win or lose, you will return to the classroom after finishing the game, and will need to speak to both Natsai and Professor Ronen.

✧ Defence Against the Dark Arts Class Walkthrough

Defense Against the Dark Arts Class is the fourth main quest within Hogwarts Legacy; where you continue your training on your first day by learning the Levioso spell.

Beginning the Lesson

After tracking Defense Against the Dark Arts Class, head to the Defense Against the Dark Arts classroom. Once you enter and begin the lesson, you'll meet Professor Hecat, Sebastian Sallow, and Leander Prewett. Here you will learn Levioso, completing the same spell-learning minigame as before. Hecat will then instruct you to perform the same on the dummy. Use a Basic Cast, then use Levioso (to break the Yellow shield), followed by a Basic Cast to complete the tutorial.

Dueling Sebastian Sallow

You will then begin a duel against Sebastian Sallow, the first one in the game. Here you can use Basic Cast, Levioso, and Protego (and the Stupefy counter-attack). Win or lose the duel, Hecat will congratulate you on your effort, and you'll then need to speak to Sebastian. He'll inform you of a secret dueling club, led by Lucan Brattleby near the Clocktower.

✧ Weasley After Class

Weasley After Class is the fifth main quest within Hogwarts Legacy; where you prepare to depart for Hogsmeade, but not before learning a new spell, Reparo, whilst completing additional assignments from Professor Ronen.

Meet Professor Weasley

Head to the Transfiguration classroom to begin, where you will be introduced to the House Elf, Deek. Professor Weasley will discuss the additional assignments that Professors will give to you to help you catch up to other students, as well as discuss the imminent trip to Hogsmeade to refresh your supplies.

Professor Weasley suggests traveling to Hogsmeade with a fellow student, and you get to choose between Natsai Onai and Sebastian Sallow. Whomever you choose, you are informed to speak with Professor Ronen about an additional assignment.

Professor Ronen's Assignment - Learn Reparo

Report to Professor Ronen in the courtyard outside, and he'll set you some tasks prior to learning Reparo. The first is as follows:

- Collect the flying page near the broken statue

- Collect the flying page in the Defense Against the Dark Arts Tower

You can collect these pages by looking at them as they fly around and casting Accio. Track their general location using the mini-map, then keep your eyes up to see a book flapping around in the air within that general area. With them both collected, return to Ronen and you'll need to complete the spell-learning minigame as before, this time in the shape of a spiral.

If you wish to practice, you can cast the spell on the broken statue near the body of water on one side of the courtyard.

✧ Welcome to Hogsmeade

Welcome to Hogsmeade is the sixth main quest within Hogwarts Legacy; where you venture into Hogsmeade for the first time, restocking on supplies, before fending off a surprise attack from deadly enemies.

Departing for Hogsmeade

After choosing your traveling companion to go to Hogsmeade with, meet either Natsai or Sebastian at their respective locations. Follow them outside and then out of Hogwarts grounds, down the road towards Hogsmeade. They will point out bushes of Lacewing flies on the way that you can collect for potion-making if you wish.

Continue down the path and you will eventually meet Mr Moon, Hogwarts caretaker, who drunkenly warns you of Demiguises. Keep moving and you will eventually arrive in Hogsmeade, where you will be left to your own devices to explore. Most buildings will have locks on for the time being.

In any order, you can visit the following shops:

- Tomes and Scrolls

- Ollivanders

- J. Pippin's Potions

- The Magic Neep

Tomes and Scrolls

At the very south end of Hogsmeade, you'll find Tomes and Scrolls. Head inside and speak to Thomas Brown. After a short conversation, you can purchase the Potting Table with a Small Pot and the Small Potions Station Spellcrafts for free.

Ollivanders

Just opposite The Three Broomsticks in the center of Hogsmeade, you'll find Ollivanders. Head inside and use the bell on the counter to speak with Gerbold Ollivander. After a failed attempt or two, a wand reveals itself. Here you can customize your wand to your liking, and once completed you and the wand will synchronize perfectly.

J. Pippin's Potions

Located near the West Hogsmeade Floo Flame, head inside and speak to Parry Pippin. After a short conversation, you can purchase the Edurus Potion Recipe and the Wiggenweld Potion Recipe for free.

The Magic Neep

Located just northwest of J. Pippin's Potions, speak to Timothy Teasdale outside The Magic Neep. After a short conversation, you can purchase a packet of Dittany seeds for free.

The Troll Attack

With all 4 shops visited, head to meet Natsai/Sebastian in the town circle. As you do, a pair of Armored Mountain Trolls attack the village. Whilst residents lead one away, you and your friend need to deal with the other. Here you learn to dodge, use the Ancient Magic Throw, and Ancient Magic Finisher.

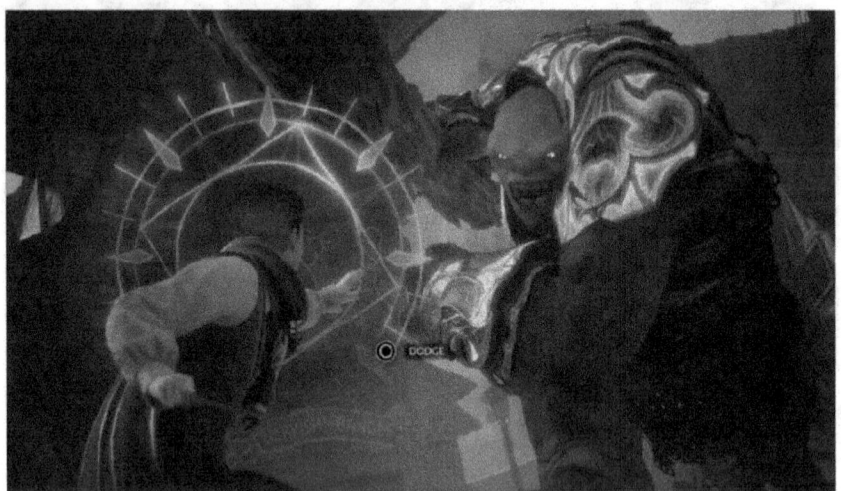

With the troll defeated, use Reparo to repair the various broken stalls/buildings/stacks around the town square, and then speak to the shopkeeper nearby in Gladrags Wizardwear. Augustus Hill will give you a set of robes to thank you for saving the village.

The Three Broomsticks

Next, you'll need to head to The Three Broomsticks tavern in Hogsmeade. As you make your way there, you will overhear a conversation between Victor Rookwood and Ranrok, seemingly scheming together.

As you head into the tavern to evade detection, you meet Sirona Ryan, the landlady. Rookwood and his lackey Theophilis walk in, and a standoff ensues. As Rookwood is forced out by Sirona, you and your friend decide it is time to return to Hogwarts.

As you exit The Three Broomsticks, you can use the Map to fast travel via Floo Flame to your common room.

Talk to Professor Fig

After reading a letter from Professor Fig you received via Owl Post, leave your common room and speak with him in his classroom in the Defense Against the Dark Arts tower. He claims he read an inscription on the locket that revealed a map.

The map highlights a forbidden section of the library, and Fig declares you need to hone your defensive magic further before he feels comfortable tracking into the dangerous and dark areas of the library.

This will lead you to the 'Professor Hecat's Assignment I' Side Quest which is essential to progress.

Professor Hecat's Assignment I - Learn Incendio

Returning to Hecat's classroom, she will help you to learn Incendio, the fire-making spell. In order to do so, she'll ask you to:

- Win 2 rounds of Crossed Wands

- Complete a round of spell combination practice with Lucan Brattleby

Both of these are completed at the Clocktower; the spell combination practice simply requires you to follow the spell prompts at the top of the screen, performing them all before the practice dummy hits the floor. Complete 3 consecutively more difficult rounds for the practice to be complete.

Then participate in 2 total rounds of Crossed Wands, defeating fellow students in spell combat. Round 1 is against two opponents, whilst Round 2 is against 3.

Once both objectives are complete, return to Professor Hecat and complete the spell-learning minigame to successfully learn and cast Incendio. With this learned, you'll have enough spells that you can swap ones out for your active Spell Set.

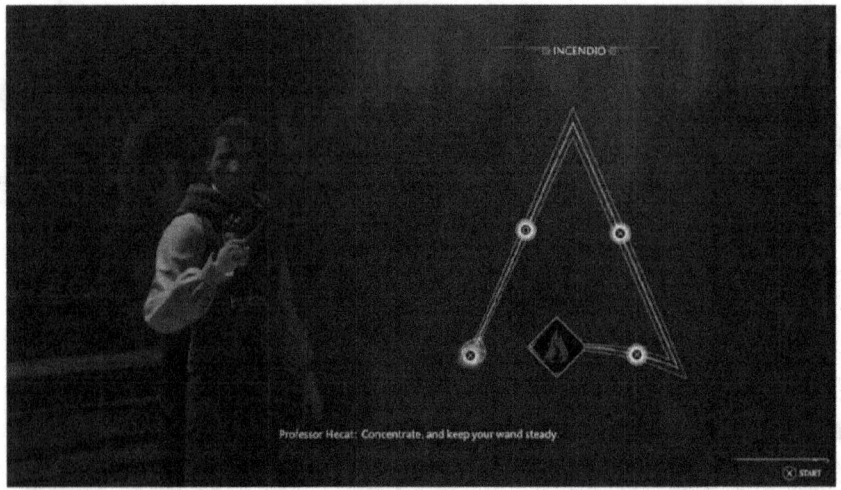

✧ Secrets of the Restricted Section

Speak with Professor Fig

After learning Incendio from Professor Hecat, return to Professor Fig in his classroom. Just before the two of you depart for the library, Headmaster Black arrives summoning Fig away. With the trip delayed, you must find and speak with Sebastian Sallow, who supposedly knows a way into the Forbidden Section.

Speak with Sebastian

Talking to Sebastian, you explain the situation and he agrees to keep the secret and help you out. He asks you to meet him outside the library later that night. Do so, meeting him in Central Hall, and he'll point out the door you need to speak to.

Learning Disillusionment

In order to reach there undetected, he will teach you the Disillusionment charm via another spell-learning minigame. Successfully complete it to learn and cast Disillusionment, turning yourself invisible.

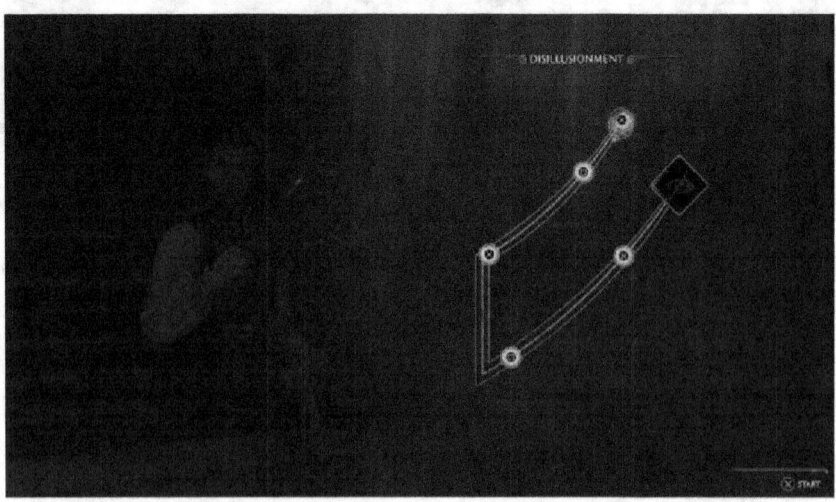

Assign it to a slot, cast it, and descend down the steps. Using the tutorials, be sure to avoid detection from the prefects roaming the halls.

As you enter through the door, down the stairs into the library, you'll need to take cover behind the bookcase on the left.

After the cutscene, wait for Sebastian to cause a distraction, luring the librarian away, before moving forward and looting her desk for the key. Then proceed to the locked Restricted Section gate and unlock it.

The Restricted Section

Move forward down the steps to receive a tutorial on distracting enemies. Use your Basic Cast and your aiming function to hit the armor stand to the left of the bottom of the stairs to lure the first ghost over, then loop around the opposite side to proceed. Carefully maneuver around ahead to evade a second ghost, then go down the stairs beyond.

Once at the bottom, you're safe from detection and can move freely. As you descend into the storage area you can use Levioso to get a Field Guide page from the statue, and also loot a number of small chests for a wand handle and some Gold.

As you round a corner, Peeves will appear and promise to rat the both of you out. Sebastian pursues Peeves, whilst you continue alone. Use Reparo to lift the armor blocking the path, then push forward down the steps. At the bottom you'll enter a circular room with an Ancient Magic spot on the floor. Before interacting with it, loot the large chest down the steps for a cosmetic item.

Upon interacting with the Ancient Magic spot, the portal ahead will activate, allowing you to enter the Athenaeum. Head down the steps, looting the large chest on the left at the bottom, before interacting with the doors ahead.

The Athenaeum

Once inside, head forward and use your Basic Cast on the glowing orb above the archway ahead of you to raise a bridge. Cross over and you'll face off against 4 Pensieve Sentries. Here you learn about Ancient Magic accumulation through combos and successful attacks.

10+ hit combos will begin generating Ancient Magic Powerups, which will greatly fill your Ancient Magic Meter and recover a small amount of your Health. As you fill up one of the bars of your Ancient Magic Meter, you can cast devastating attacks that break Shield Charms and deal massive damage.

After defeating them, the next door will open and you can proceed through. In the next room, there are 2 Pensieve Sentinels, which can perform ranged attacks.

With them defeated, you can use Basic Cast on the glowing orb above the doorway (as before), and you'll want to run and jump across both ends of the bridge that arises, as the timer is very brief before it drops back down.

NOTE: There is also a chest in this same room in an alcove to the left that seems inaccessible; you can in fact reach it by walking near the flat jutting out section on the initial platform in that room. A bridge will rise as you walk that will lead you to the chest.

In the next room down the corridor, you can reach another chest by hitting the glowing orb with Basic Cast and using a secret bridge on the right side of the platform. Once you reach the wall where the bridge stops rising, hit the orb again to raise the remainder, letting you reach a chest in an alcove on the right wall.

To progress properly, you'll need to hit the glowing orb with Basic Cast so the first half of the bridge (leading towards the orb) is on your side. Walk to the end, then hit the orb again to raise the other half. Be sure to quickly transition from one to the other before the first half you were on drops away.

There is a chest to both your left and right before you proceed into the final room. You'll note a number of both Pensieve Sentries and Pensieve Sentinels.

Defeat them all as you see fit, and move forward to trigger a cutscene where you receive another Pensieve memory from a book, where the witches and wizards of the past use their magic to bring water and vegetation to a starving village, with a young girl, Isidora Morganach, watching on and later becoming a student at Hogwarts as a 5th year, just like you.

You return to the main library just in time to see Sebastian being reprimanded by Peeves and the Librarian; yet he doesn't rat you out and takes the blame fully upon himself.

✧ Tomes and Tribulations

Return to Professor Fig

With the excursion to the Forbidden Section of the library over, you need to return to Professor Fig's classroom and speak with him. You will interrupt a conversation between Fig and Professor Sharp, and after Sharp leaves you can give Fig the book. Fig must depart for the Ministry to explain George Osric's death in person, and will search for information about the book while doing so. You, on the other hand, are tasked with finding out about the missing pages of the book.

✧ Herbology Class

Arriving at the Greenhouses

Proceed into the Greenhouses area and enter the classroom on the left to be introduced to Professor Garlick, your Herbology teacher. She will introduce you to the class, before beginning a lecture on the Mandrake root.

Dealing with Plants

Pull out the Mandrake from the pot, then replant it following the on-screen instructions. Next, speak with Professor Garlick privately in the corner of the room, and then walk over to your potting table within the classroom.

Plant the Dittany seeds in the pot, and Garlick will congratulate you. She will then introduce you to the Chinese Chomping Cabbage. Meet Leander Prewett who will follow you as you head to the other greenhouse room and pick up the Chinese Chomping Cabbage from their flowerbeds. Select and equip them in your Tool Menu, and then use them to attack the practice dummy at the end of the room.

After speaking to Leander about the usefulness of the cabbages, return to class and speak with Professor Garlick once again. After a short conversation, the quest will conclude.

✧ Potions Class

Attend Potions Class

Head to the Potions classroom and begin the quest, being introduced formally to Professor Sharp. The first challenge is to brew a Wiggenweld Potion.

Brewing Wiggenweld

First, follow the on-screen instructions to crush up your ingredients for the potion. Then add a pinch of Dittany with the button prompt, and the subsequent drops of Horklump Juice. Similar to the grinding, stir the cauldron, and the potion will be complete.

Brewing Edurus

Professor Sharp will congratulate you, and talk about the Edurus Potion you picked up the recipe for in Hogsmeade. He tells you to collect the following ingredients, both of which can be found in his office:

- Ashwinder Eggs

- Dark Mongrel Fur

As you approach his office in the classroom, Garreth Weasley will interrupt you. He proposes that you take the opportunity to steal an additional ingredient, a Fwooper feather, for his own brew. If you choose to help Garreth, collect the feather and deliver it to him. With the other ingredients collected, head to your Potions Station in the classroom and begin brewing.

Select Edurus and wait for the timer to hit 0. Garreth's potion will explode in the meantime. After the timer on your Edurus Potion hits 0, collect it and then speak to Professor Sharp. If you help Garreth, he will chastise you, but owning up to it will cause him to let you off the hook.

After he congratulates you on your brewing efforts, the quest will conclude.

✧ The Girl From Uagadou

Meeting Natsai at Lower Hogsfield

Upon meeting Natsai in Lower Hogsfield, she will lament she hasn't been proactive like you up until now. She suggests taking down Theophilis to cripple Rookwood and Ranroks operation. She will tell you to wait until she has collated more information, and as she leaves you hear someone cry out for help, leading directly into the Trials of Merlin quest.

✧ Trials of Merlin

Help the Stranger

As Natsai departs from Lower Hogsfield, a stranger cries out in the distance. Head to the tent near the lake to encounter Nora Treadwell being confronted by two Ashwinder Scouts. Defeat them and two more will arrive, alongside an Ashwinder Duellist.

Duelling Feat

This will also be the point in which you receive your first Duelling Feat of the game, where you must complete the battle in under 1 minute. This is best achieved by using various spells in a combo. Accio an enemy towards you, Incendio them, hit them with some Basic Cast shots, then Levioso them to keep them disabled and get in more Basic Cast hits. The Stupefy counter-attack is also very useful here. When the Ashwinder Duellist apparates in, make them your priority. Defeating them takes a powerful enemy off of the field, letting you deal with the Ashwinder Scouts more efficiently and safely. Successfully defeat all 5 enemies in under 1 minute to earn additional XP and progress towards your Duelling Feats Challenges.

The First Merlin Trial

Speaking to Nora reveals the Ashwinders work with Rookwood, and she also explains she is out here doing research into Merlin, the ancient wizard. She details how Merlin was once a student at Hogwarts, and that he set up various 'trials' around the area. This is your introduction to the Merlin Trials available within the game.

She explains the Mallowsweet herb is most likely the key to initiate the trials, and so she asks you to retrieve some from the chest by her tent. Do so, and then go to place some Mallowsweet leaves in the center of the stone swirl on the ground.

The vines around the pillars nearby will disappear, allowing you to Incendio the 3 braziers atop each one.

This will solve the first trial, and to complete the quest simply speak to Nora once more. Completing Merlin Trials are an excellent thing to do as early on as possible from this point, as the Challenge rewards for dong them is expanded gear capacity, meaning you have to destroy/sell/manage your inventory far less often!

Ollivander's Heirloom (Ravenclaw)

❖ Meet with Ollivander

Head to Hogsmeade and speak with Gerbold Ollivander who requests your assistance in tracking down a special wand that has been missing for a long time. A man named Jackdaw took it, and Ollivander believes there may be a clue somewhere near the Owlery.

❖ Investigating the Owlery

Head to the Owlery and walk up to the top. You'll note among the small alcoves where owls are resting are little rings that you can use Accio to pull down. Behind some will be nothing, others will have small bags of Gold, and 3 will contain a small Jackdaw statue. Be vigilant, one of these alcoves you'll need to pull down is up quite high.

Once you have pulled down all the alcoves and collected 3 Jackdaw statues, place all 3 in the spot you are prompted to beneath the already visible Jackdaw statue, then climb up the ladder and again pull down the ringed plates covering the alcoves to cause more Gold bags and 2 additional Jackdaw statues.

Place those 2 with the others to trigger a cutscene where you are carried to the top of the Owlery by a gust of wind. At the top, you meet the ghost of Richard Jackdaw. Richard promises to help you retrieve both the wand and the missing pages of the book should you meet him outside the Forbidden Forest.

The Hunt for the Missing Pages (Gryffindor)

❖ Meet Nearly Headless Nick

You will receive an Owl Post letter from Nellie Oggspire telling you that Nearly Headless Nick wishes to meet with you. Just outside the Great Hall, by the House Point Hourglasses, is where you'll find Nearly Headless Nick.

Nick explains that he knows about the book you found, and the missing pages within. Follow Nick and he will explain that he can give you information, so long as you go into the kitchens and find some rotten roast beef for him.

Down the steps near the spiral staircase, Nick will tell you to tickle the pear in the nearby painting. Do so to open up the secret door to the House Elf kitchens. At the very back of the room across from where you entered, you'll find a small alcove with some rotten roast beef on top of the barrel on the right.

In a cutscene, the House Elf Feenky appears and agrees to let you take the rotten meat. Exit the kitchen and speak to Nick once more and he says to get the meat to Sir Patrick so that you can engage in the Headless Hunt and speak to Richard Jackdaw by proxy, who happens to have information about the book.

❖ (Almost) The Headless Hunt

Arriving in the cemetery outside Hogwarts, Nick introduces you to Sir Patrick Delaney-Podmore, and you also meet the headless ghost of Jackdaw. Without a head, you can't speak with Jackdaw, so you need to find his head.

Though Nick's attempts to join the hunt are unsuccessful, you can head just outside the cemetery to speak to the ghost of Sir Nestor Amset. He gives consent to speak to Richard Jackdaw, but you need to find his head amongst the pumpkins a total of 5 times.

❖ Getting A Head Of The Game

Agreeing to the game, walk around the perimeter of The Magic Neep and smash pumpkins with your Basic Cast. You'll know the pumpkin that Jackdaw is currently hiding in as it will slightly bulge and bounce as you get close.

You can also try to vaguely track Jackdaw's ghost as it floats away from the pumpkin you destroyed, and it will give you a sense of the general area he is now hiding in. Find him 5 times and you can then speak to him, requesting help in finding the missing pages.

Jackdaw explains he stole them from Peeves a long time ago, and that your best bet is to meet him just outside the Forbidden Forest in order to track them down. Agreeing to meet him there, the quest concludes.

Prisoner of Love (Hufflepuff)

❖ Meet Eldritch Diggory

You will receive an Owl Post letter from Lenora Everleigh stating that a portrait in the Hufflepuff common room wishes to speak with you, Eldritch Diggory (the former Minister of Magic) to be precise.

Head to the Hufflepuff common room and speak with the portrait to the left of the entrance/exit. Eldritch explains that you can likely help him solve a decades-old murder mystery, and in exchange he can help you find the missing pages from the book you've found. Someone went missing, and someone was charged with their murder with no evidence, and those missing pages could exonerate them.

He tells you his great niece Helen Thistlewood can give you extra details, and you'll need to speak to her in Upper Hogsfield.

❖ Speak to Helen Thistlewood

Heading to Upper Hogsfield, speak to Helen in her cottage and she explains a boy named Richard Jackdaw went missing going to meet a girl named Anne. His headless ghost was spotted shortly thereafter. Due to a ludicrous alibi involving Jackdaw stealing pages of the book from Peeves, and a woman named Apollonia Black giving potentially false testimony, Anne was condemned to life in Azkaban.

Helen explains that Anne, though a shell of her former self, has most likely deciphered some of the puzzles that Jackdaw laid out for her before his death, and that you should speak with her to see what can gleaned.

❖ Arriving in Azkaban

Agreeing to go with Helen, you both Apparate straight to Azkaban. Almost immediately, a group of Dementors descend upon you and Helen protects you both with Expecto Patronum.

Proceed down the hallway and speak to Anne in her cell, who reveals that Jackdaw hid something in a vault within some ruins near Upper Hogsfield. Before you depart, Anne swipes at Helen, injuring her, before you both Apparate back to Upper Hogsfield.

Helen, wounded, tells you to go on before Apparating away again.

❖ Finding What Jackdaw Left Behind For Anne

Proceed down the hill ahead of you, past the dam, to the entrance of a small cave.

Once inside, you'll arrive in a small room with a number of wrungs you can pull on with Accio. Only a specific few will remain out once pulled, and an incorrect interaction will reset them all. Use Revelio to turn certain ones blue, denoting the ones you need to pull out with Accio to open the gateway ahead.

In the now accessible room, in the immediate left corner, is a Jackdaw's Clue scroll on a pedestal. With the scroll collected, leave the vault and you'll immediately encounter Richard Jackdaw's ghost. You both catch each other up on how you came to be here and what became of Anne. Jackdaw says he must speak with Helen to discuss Anne and the actualities of his disappearance, potentially to clear her name.

Before he leaves, he says he can help you get the pages of the book back, as they are likely still with his body, somewhere deep within the Forbidden Forest. As the conversation ends, the quest concludes.

Scrope's Last Hope (Slytherin)

❖ Find Scrope's Note

You'll receive an Owl Post letter from Scrope the House Elf telling you that he knows about the book you've retrieved, and has additional information. He says to find his note in the courtyard near the clocktower.

Head to the Clock Tower Courtyard Floo Flame location and then proceed outside to find a Mysterious Note in the Gryffin statues mouth on the corner of the central well. The note explains it isn't safe to speak in public, so cross over the wooden bridge to find the next note.

The note will be up high on a rock to the left, so you'll need to Accio one of the nearby crates over so you can climb up and reach the note. This second note tells you to search the pumpkins just down the hill for a final third note.

You'll need to break the pumpkins by using your Basic Cast on them, eventually revealing the third note inside one of the larger pumpkins round the right side of the building.

❖ Meet Scrope

With all three notes found, head down to the coast near the broken docks to meet Scrope in person. Scrope explains he is the headmasters House Elf, and could be in real trouble should he be spotted.

Scrope explains his former mistress Apollonia Black knew of the pages being torn from the book, and the nearby grotto of Apollonia's could hold a clue. He also asks that whilst you search for the pages you also collect a ring that he can give as a gift to the headmaster.

❖ Finding the Grotto

He explains you'll need to place a slice of toast on a pedestal within the grotto, and wishes you luck. Continue along the coastline towards the cliff face directly beneath Hogwarts to find a small cave that you'll need to swim into.

Inside, youll find a mural of the giant squid that occupies the lake, as well as the pedestal. Interact with the pedestal for the mural to come to life, accept the toast, and then create a passageway.

❖ Inside the Grotto

On the table inside the grotto you'll find Apollonia Black's journal, and upon doing so the headless ghost of Richard Jackdaw emerges from the cave walls. He explains he sold the ring a long time ago, and that he stole the pages from the book to try and impress Apollonia. Asking him for the pages, he explains they are actually with his corpse somewhere deep within the Forbidden Forest, as he tried to follow the map himself. You agree to meet him promptly at the edge of the Forbidden Forest.

❖ Return to Scrope

Returning to Scrope back down the coastline, you inform him that your search for the pages has beared some fruit, but alas Scrope's hope for the ring was futile. He thanks you for your efforts regardless and Apparates away, as the quest concludes.

⬦ Jackdaw's Rest

Meet Richard Jackdaw

Head to the Floo Flame location just outside the Forbidden Forest and travel into the Forest itself to meet up with Richard Jackdaw. Speak to him, and then follow him to the East North Hogwarts Floo Flame spot. From here, you'll need to progress alone deeper into the forest, following the waypoint on your minimap. Eventually you will arrive at the Jackdaw's Tomb Floo Flame spot, and notice the birdbath in the center of the area (just as Richard described).

Approach it and interact to whisper the password which will open up the swirling stone portal ahead of you. Before you can enter though, three Goblins ambush you. Defeat them (with the Duelling Feat being to break a Shield with an Ancient Magic Throw), and then proceed inside the tomb.

Inside Jackdaw's Tomb

Once inside the tomb, proceed forward to the locked door. You'll see 3 orbs with the Ancient Magic symbol. Hit all 3 with a Basic Cast in quick succession to unlock the door and create a magical bridge beyond.

Through the door, you'll come to a spider web. Burn it with Incendio, and you'll quickly be ambushed by various Thornback Spider enemies. For a Duelling Feat, hit another Spider with a Spider explosion. To do this, hit a spider with Incendio, then use Basic Cast on them whilst they are alight (and near another Spider) to cause an explosion that deals collateral damage.

With the spiders dead, you can continue down to the left, or crouch through a small gap to the right. If you go through the gap to the right, you can face more Thornback Spiders (Duelling Feat being to lift a burrowing Scurrior with Levioso, and to simply defeat all enemies), and afterwards you can loot the large chest on the far side of the room.

Getting back on the main path down into the caves, you can turn right at the bottom to find a small platforming puzzle. Go up the ramp to the right and use Accio to summon the platform. Then aim to the far side and use Accio again to pull the platform you're on back across the room, enabling you to loot a large chest around the corner.

Continuing down the main path, you'll be ambushed by spiders from above as you pass through a small and narrow ravine. On the other side, you'll find another locked door with orbs you'll need to hit in quick succession. The third orb is tucked away up and to the left of the other two, so aim for that one first.

Successfully hitting all 3 quickly will cause the door to open and another bridge to form. Continue ahead and you'll reach a platforming puzzle section. Use Accio to summon the platform to you, and once on it you can use Accio again to pull yourself either directly ahead, left, or right. Going right will net you a small chest and then a larger one should you push through the cave ahead full of spider eggs.

Going left allows you to reach a single small chest. Ultimately, you'll want to go directly ahead, which allows you to traverse a number of different passageways that include Accio platforming sections to reach various high up areas with loot chests.

If you simply wish to progress, follow the initial path round and drop down into a small pool of water.

Shortly thereafter, you'll notice the second bridge you assembled, and will be ambushed by a large number of Spiders. (Duelling Feat is to defeat a Spider with Ancient Magic, and to keep an enemy airborne for 10 seconds - to do this, simply cast Accio on a spider, then at the last second cast Levioso, then hit Accio again as soon as it recharges, and repeat).

After the first wave of spiders are defeated, two Thornback Matriarchs will appear with additional spider backup.

Clear these enemies out, then be ready to hit another 3 orbs in quick succession with your Basic Cast. One is down the slope ahead, and the other two are up top either side of the archway. Stand as far up the slope as you can before you're unable to hit the orb at the bottom, then try and hit all 3 quickly (this one does seem to have a longer window for errors/gaps between hits).

Jackdaw's Skeleton

After successfully doing so, the bridge braziers will light, and the doors beyond will open. Cross over the bridge to find Jackdaw's skeleton, and though Ollivander's wand is shattered, you can still recover the pages of the book from his body. Do so, and the Pensieve Sentries and Sentinels lining the room will activate.

After defeating them all, the large Pensieve Protector at the back of the room will activate. You must defeat it, and the other two will also activate, but now at the same time. Defeat the two of them, then head to the archway that appears at the back of the room.

You'll find another Ancient Magic spot on the ground, which you can interact with to open the archway portal. Walk forward down the corridor and the room will begin to flood. Your Ancient Magic will automatically protect you, so keep moving forwards through the next archway and up the steps to arrive in a spacious room with 4 golden arches on the wall that each reveal a different animated portrait.

Speak to Percival Rackham through the portrait and he'll recognize that someone has finally found his map chamber.

After a brief conversation, you gain access to the Talents system within Hogwarts Legacy, acting as Skill/Talent Points to spend on upgrading/unlocking certain abilities. You begin earning Talent Points from Level 5 onwards, and as the max level is 40, you will NOT be able to unlock all Talents within the game, so choose wisely which abilities/paths you want to invest in!

After spending your available points on whatever you like, simply exit the Map Chamber through the door behind you, then walk down the corridor and up the spiral staircase to arrive back in the lower levels of Hogwarts.

✧ Flying Class

Attending Flying Class

To attend Flying class, head outside into the north courtyard of Hogwarts and begin the class near the broom stacks. Here you will meet Madam Kogawa, your Flying professor.

After a brief cutscene, complete the on screen prompts to summon your broom up into your hand.

Now in control in the air, guide your broom through the 3 large rings in front of you. After successfully doing so, you'll be introduced to the boost on your broom, which you'll need to use to speed through various rings as you fly around the entirety of Hogwarts.

As you pass through the final one, Everett Clopton of Ravenclaw asks if you're interested in some fun. Regardless of how you answer, you'll need to follow him around Hogwarts once again, this time being told to lean forward into an additional speed boost. Land back down in the initial courtyard, and then Everett will suggest you visit Albie Weekes' shop in Hogsmeade to purchase your own broom.

✧ The Room of Requirement

Meet Professor Weasley

Head up to near the top of the Astronomy Tower to meet with Professor Weasley, and a hidden door in the wall will reveal itself in a cutscene.

Follow Professor Weasley through the maze of items and she will eventually teach you Evanesco, the vanishing spell, when you reach a barricade of stacked chairs. Once you've completed the spell-learning minigame, cast Evanesco on the stacked chairs and continue following Professor Weasley.

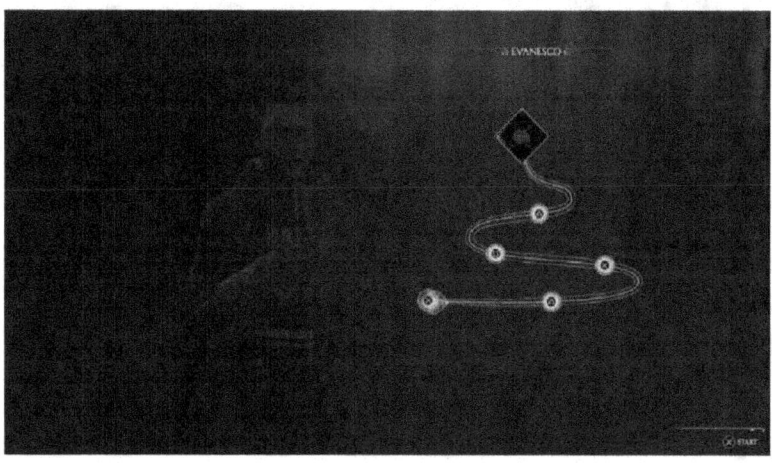

Exploring the Room of Requirement

As she finds her old school bag, you can continue forward on your own. You'll see a container tucked away to the left after clearing the second stack of chairs with Evanesco. Pull the container out with Accio, then lift it with Levioso to climb on it as it floats, which lets you reach the chest above.

Continuing on, clear a third stack of chairs with Evanesco, and crouch through the gap, proceeding down the path clearing stacks of chairs as you go. Eventually you will meet with the House Elf, Deek, once again. He will inform you that you can imagine this room to be whatever you need, and you watch in a cutscene as the room transforms before your eyes.

Head to the Desk of Description in the corner of the room to identify any/all unidentified pieces of equipment in your inventory.

With that done, speak to Professor Weasley. She'll give you a lesson in Conjuring, beginning with the Conjuring Spell. More complex objects you can summon will require a Spellcraft, which can be discovered or purchased in Hogsmeade.

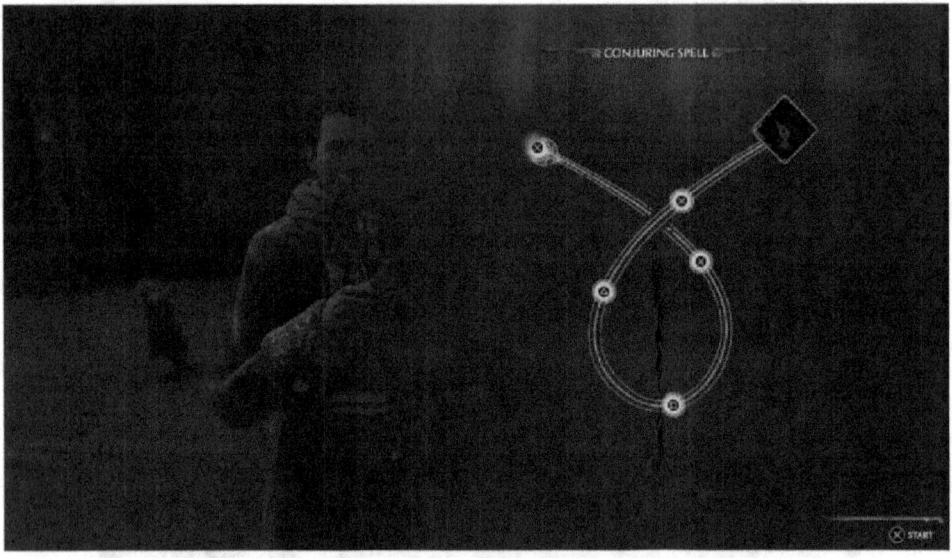

Conjuring Your Stations

You'll now need to summon a Potting Table and a Potions Station. Assign the Conjuring Spell to your quick select menu, and cast it. Select the Potions Station and the Potting Table from the Potions and Herbology tabs respectively.

As you prepare to place them wherever you like in the room, you can alter their visual style and color. After placing both stations down, you'll briefly speak to Professor Weasley once more before the quest concludes.

◇ In the Shadow of the Undercroft

Meeting Sebastian

Meet Sebastian at night in the Defense Against the Dark Arts Tower, and after a brief conversation where you thank him for his efforts in the library, follow him down the corridor towards a dead end, and interact with the cabinet on the right to use the secret entrance to the Undercroft.

Inside the Undercroft

Once in the Undercroft, you can loot the various chests in the room as Sebastian talks, and then go to him and speak, and he requests you keep the existence and location of the Undercroft a secret. He also explains it is a good place to practice forbidden spells, like the Blasting Curse, Confringo, which he goes on to teach you.

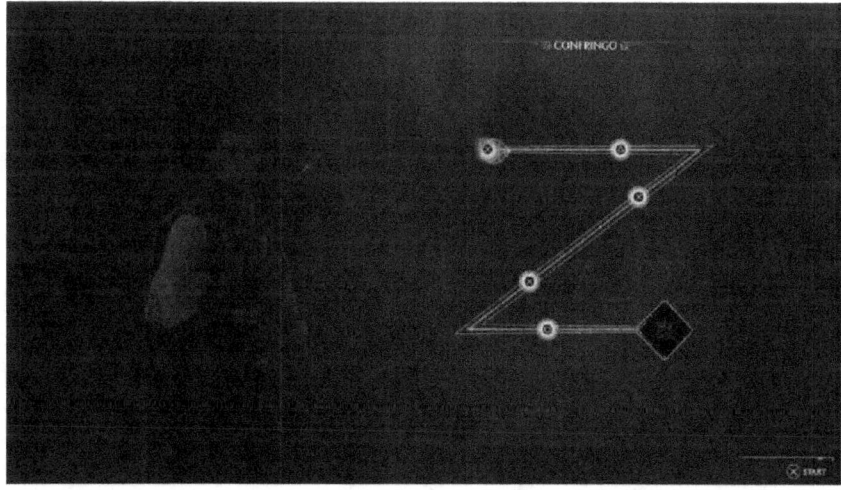

Learning Confringo

Successfully complete the spell-learning minigame in order to learn and then cast Confringo. Proceed around the room hitting the 3 unlit chandeliers in the room with Confringo, then report back to Sebastian.

After the conversation concludes, leave the Undercroft to run into Ominis Gaunt who threatens to have you expelled if you talk about the Undercroft.

✦ The Map Chamber

Reunite with Fig

With Fig returning from the Ministry, meet him at the back of his classroom. You catch him up on what has happened since he's been away, and will then teleport to near the Map Chamber. Head forward and down the stairs to reach the Map Chamber, and Fig will be enthralled with what he sees.

After a magical light appears on the podium, Fig places the book down to reveal a spectral map on the floor in the center of the room.

Speak to Rackham

Interact with the portrait on the wall to begin a conversation with Percival Rackham, who will subsequently reveal that 4 trials will appear on the map that you must complete to truly master your Ancient Magic abilities.

Though Rackham initially tries to postpone the trials beginning, Fig explains the Dark Magic that the Goblins are beginning to possess, so Rackham reluctantly agrees to begin the first trial.

✦ Percival Rackham's Trial

Investigate the Goblin Presence

Meet Professor Fig at the San Bakar's Tower Floo Flame spot to the far north of Hogsmeade, and once in control again after a brief conversation, cast your Disillusionment spell and head towards the tower.

This is where you get tutorialized on the use of Petrificus Totalus, the sneak attack insta-takedown that you've likely already used by this point. Simply sneak up behind an enemy whilst using Disillusionment and hit the prompt that appears.

Enter and deal with as many or as few Goblins as you'd like, then turn to the tent on the left after you enter the ruins to find a letter with orders from Ranrok resting on a desk. As you leave the tent, a Loyalist Assassin will appear, as well as additional Goblin soldiers that you'll need to defeat.

With the enemies defeated, head to the drawbridge on the opposite side to the tent, and Fig will unlock the doors, allowing you to climb the spiral staircase inside. At the very top of the stairs you'll find a portrait of Rackham. Speak to it to continue.

Rackham will explain that downstairs a reservoir of Ancient Magic has been activated, and you need to investigate it. Heading back downstairs, you'll find the Ancient Magic spot just inside to the left from the entryway to the tower. Investigate it to open up an archway to Percival Rackham's trial.

Rackham's Trial

Passing through the magical archway, continue down the corridor and the following steps to reach a large door. Enter to begin the trial.

Walk all the way down to the end, and then down the staircase on either side to find another Ancient Magic spot to investigate.

Do so to create a bridge spanning the gap above you. Go back up the stairs and cross the bridge to the other side, looting the small chests to the left and right if you wish.

Continue into the next room and across to the other side of the archway to find yet another Ancient Magic spot. Investigate it and the archway behind you will now have a portal leading you down a brand new corridor.

In the room at the end, on the far side, a Pensieve Protector and two Pensieve Sentinels will awaken. Defeat them however you see fit and then proceed into the next room.

You'll need Accio equipped for this puzzle section. Use it to summon the floating platform towards you, climb aboard, then use Accio to pull yourself towards the column in the center of the room.

From there, look right and you can Accio again to reach the platform and investigate another Ancient Magic spot.

Doing so creates a portal in the archway to your left, letting you pass through it to loot a large chest.

Come back out, and then climb back on the floating platform. With additional rings to use Accio on throughout the room, you can pull yourself to the central column, and then the far side of the room to loot two small chests, then turn around and use Accio to pull yourself to the wrung on the opposite side of the archway portal, letting you access the room ahead.

In there, two more Pensieve Protectors will awaken, and once they are dealt with another puzzle platforming section arises. Walk up the steps to the right of the room, and then turn around to pull the floating platform with Accio towards the wisps of Ancient Magic, letting you climb on and then investigate the magic.

Drop down and proceed through the archway portal on the blue side, and you'll notice the walkway you were just on has disappeared, allowing you to go back up the steps, climb on the floating platform, and freely Accio yourself over to the distant platform with a chest.

With that done, drop down and go back through the portal to bring the walkway back, go up and then Accio the floating platform over to yourself, letting you progress across (as the other Accio ring pull has returned on this side of the portal).

Continue down the large spacious corridor into the next room, where two more Pensieve Protectors await as well as five Pensieve Sentinels. The Sentinels should be your first priority, as they are easier to take down than the Protectors, and their ranged attacks can easily take you out if you are focused on dealing with/dodging the Protectors.

After the enemies are dealt with, go down the steps to the right of this space, and interact with yet another Ancient Magic spot. You'll note the archway portal in the distance.

Accio the floating platform toward you and climb on. You can then Accio yourself towards (and past, if you wish, to loot the large chest behind) the archway, and then turn to your left (facing the room from the way you came in) and begin to use Accio on the far ring, but let go as you line up with the portal.

Through the portal, you'll be able to see another floating platform. Accio the platform, and that platform and the one you are on will meet in the middle of the portal. This allows you to cross over through the portal directly onto the other floating platform, which you can then use Accio with to drag yourself towards the next section of the trial.

Proceeding up the steps at the end of the corridor, and through the doors once they open, you can loot the large chest on the left and continue out across the bridge to a large circular platform.

As you arrive there, two Pensieve Protectors and four Pensieve Sentinels will drop down that you must defeat. Again prioritize the Sentinels first for their ranged attacks and easier difficulty to remove from the battlefield faster.

Pensieve Guardian Boss Fight

With all those enemies down, a gigantic Pensieve Guardian will emerge from the floor. Your Duelling Feats here will be to hit it with Ancient Magic, and to destroy the orb that it will occasionally summon as it is charging up. In order to destroy the orb, you must hit it with a spell of the corresponding color.

Ancient Magic will be your best friend in this fight, be sure to collect any Ancient Magic orbs that drop as you combo, and take full advantage of the downtime when the Guardian drops to one knee, as it will take increased damage and will drop more orbs. Once the Guardian has been defeated, cross over the final bridge to learn the trial's secret.

You'll enter a room with a gigantic statue and a Pensieve in front of it. You can loot the large chest just to the right of the Pensieve before interacting with it if you wish. Once ready, interact with the Pensieve to view yet another of Rackham's memories; this one being a training session between Rackham and Isidora Morganach, in which she laments the fact that she has yet to help her ailing father, despite her obvious magical gifts. A jump to the next memory shows Isidora older, now as a Defense Against the Dark Arts Professor.

To depart back to the Map Chamber, head to the crystallized stone archway to your left and interact with it. Then walk up to Professor Fig and Rackham's portrait to meet Charles Rookwood's portrait. Rackham speaks to Rookwood about the looming Goblin threat, and Charles decides to halt the trials until more is learned about the Goblins schemes.

You inform Fig of what you saw in the Pensieve, and you decide the best first place to look for information is with Sinora at The Three Broomsticks in Hogsmeade. As Fig departs, speak to Charles Rookwood. He informs you of Ancient Magic spots out in the open world that you can use to enhance your magic, and the quest concludes as the season of the school year changes to fall.

✧ Beasts Class

Attending Beasts Class

Head to the Beasts Classroom in The Bell Tower Wing of Hogwarts to begin Beasts class. Here you meet Professor Howin and Poppy Sweeting of Hufflepuff (assuming you're from another house and haven't met her already).

After a brief cutscene, you acquire the Beast Petting Brush which you can use to care for various Beasts. Follow the on-screen prompts to assign the Brush to a Spell Slot. Equip and then use it on the Puffskein in front of you to brush it.

You must then repeat this process with Beast Feed. Assign it to a Spell Slot, equip and use it, and you'll feed the Puffskein. After successfully doing so, Poppy will guide you to the Kneazle pen nearby. Once back in control, brush and then feed any of the Kneazles in the pen using their assignments you gave them in your Spell Slots.

Meeting Poppy's Friend

With the class concluding, go and speak to Professor Howin, who will assign you additional assignments over time. As Howin departs, speak to Poppy as she calls you over. She will thank you for defending her and the beasts from the bullies, and will ask you follow her to meet someone in the forest.

After following her for a short while, she will call down the Hippogriff, Highwing. Follow the on-screen prompts to bow twice, and you'll then need to brush and feed her. For future reference, you'll know what any beast needs by the two diamonds beneath their name at the top of the screen. The left diamond is brushing, and the right is food. If both are green, the beast is content!

After Highwing leaves, Poppy explains she saved Highwing from poachers years back, and the fact that poachers are becoming more and more common in the area, and that they seem to be planning something. As Poppy vows to get to the bottom of it with your help, the quest concludes.

✧ The Caretaker's Lunar Lament

Meeting Gladwin Moon

Meet Gladwin Moon near the Moving Staircase/Reception Hall, and he will ask you to observe the Demiguise Statue and how it changes when day switches to night. Once it does and you regain control, go up to the statue and remove the Demiguise Moon from it.

Follow Gladwin to a nearby locked door, and he'll ask you to deal with two additional Demiguise Statues One in the Hospital Wing, and the other in the Prefect's bathroom. He will then teach you Alohomora. Complete the spell-learning minigame to successfully learn and cast Alohomora.

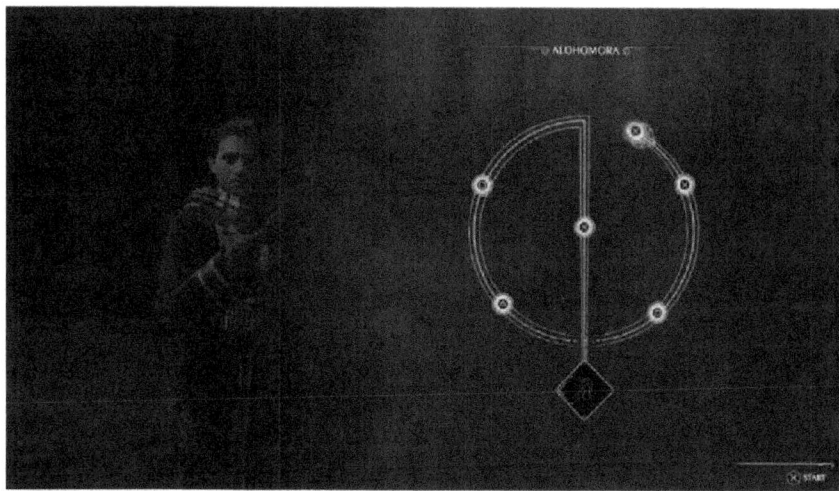

This is an Essential Spell, so will always be equipped without needing a Spell Slot. Approach the nearby locked door and cast Alohomora to begin a minigame.

Unlocking Minigame

This one can be difficult to understand to begin with. You must rotate both sets of gears until you see a reaction, and hold it in place once you do so to cause the lock to open. For example, on console, rotate the left thumbstick around in a circle until you see additional gears light up green and begin to turn.

Holding that thumbstick in that position, do the same with the other thumbstick until you see a red reaction much the same. Hold them both in place for a couple of seconds to complete the lockpicking minigame.

Finding the Demiguise Statues

After successfully doing so, cast Disillusionment and proceed through the door. Heading up the stairs, you can use Revelio to reveal nearby prefects who you'll want to avoid. On the first floor, there will be an

Arithmancy door that you can actually complete now if you wish, though you'll want to aim in a far corner and use Basic Cast to lure the prefect away.

Head up the next flight of steps and keep right to avoid the two prefects in the left corner. Head up the spiral staircase and you can access the Prefect's bathroom via a Level 1 locked door on your right.

Unlock it with the Alohomora minigame, then proceed around the right side of the bathroom to avoid the prefects and find a Demiguise statue on the far side.

Head back out and up the spiral staircase to your right to arrive at the Hospital Wing Floo Flames location. Turn left and carefully sneak past the nurse and professor near one of the beds, and you'll find the second Demiguise statue on a table at the far end of the room.

You can then carefully walk all the way back down, evading the patrolling prefects, and return the Demiguise Moons to Gladwin. Promising to help you unlock stronger forms of Alohomora if you assist him, the quest concludes.

✧ The Helm of Urtkot

Speak with Sirona

Head to The Three Broomsticks in Hogsmeade and speak with Sirona, who will tell you that they have a goblin friend called Lodgok. You mention you'd like to speak to Lodgok about Ranrok, and Sirona tells you that he'll likely be at the Hogs Head.

Speak with Lodgok

Head down the road in Hogsmeade to the Hogs Head tavern, and speak to Lodgok at the table on the left-hand side.

Explain that Sirona trust you, and Lodgok will reveal a plan of his that will allow him to learn of what Ranrok's plans are, but a sacred Goblin relic will be needed.

Meet Lodgok At the Witch's Tomb

Just outside of Hogsmeade, meet up with Lodgok and follow him to the nearby Collector's Cave Floo Flame location.

On route, he will explain the relic you are after is the Helmet of Urtkot. After arriving and having a brief conversation with Lodgok, enter the Collector's Cave.

The Collector's Cave

Once inside, head down the path and you'll arrive at a locked door with moths.

Head down the passageway to the left, and use Lumos to lure the third moth back to the door.

Stop casting Lumos to let the moth go, unlocking the door. In the room ahead, there will be a large chest in the alcove on the left.

Continuing to follow the path lit with wall sconces, you'll arrive in a circular room with another moth door on the far side. All three moths are missing from this door, so you'll need to collect them and bring them back. Two are on either side of this circular room, and the third is behind a set of broken doors, just to the right of the moth door, that you'll need to cast Depulso on to blast open.

With all three moths in place on the door, continue down the path and in the next room you'll encounter Inferi, undead enemies that can only be harmed once set on fire. Use spells like Confringo or Incendio, as well as Ancient Magic Throw using explosive barrels around the room to deal with them, with a Duelling Feat being to use a Mandrake during the fight.

Further down the path, another moth door appears before you.

In the center of the room is a mechanism, which you'll need to place one of the two moths nearby on to power it.

After doing so, step on the platform on the right side of the room, and cast Depulso on the now powered mechanism in the center to raise you up, letting you climb up the wall to reach the third moth needed for the door.

Drop down and place the moth in the door using Lumos, retrieve the moth from the mechanism, and then the other moth on ground level in this room to unlock the door and progress. In the next room down the path, you'll see a hanging crate. Use Basic Cast to break the rope and drop the crate, breaking the floor allowing you to drop down.

Before you do, you can Depulso another set of broken doors on the right, and use Accio to drag a crate around, letting you reach to high platforms in that room with one chest each.

Once ready, drop down through the new hole in the floor into the water, and proceed into the circular room ahead. You'll face a large number of Inferi here, so remember to break out Confringo and Incendio early. The Duelling Feats here will be to complete the battle in under 2 minutes, and to slice a frozen enemy (done by freezing an enemy with Glacius and then hitting them with Diffindo).

You'll note another moth door on the upper platform on the left, and a mechanism in the center back of the room.

Grab a moth from either the left or right side of the room, and use it to power up the mechanism. Walk onto the lowered section between the two higher platforms, and then hit the mechanism with Depulso to raise yourself up and grab the moth on the right side with Lumos.

Jump across the gap and put the moth in the door, then get the other moth just to the right of the door and Lumos it into the door as well. There are in fact 4 moths in total in this room, so leave the moth in the mechanism, and drop back down to grab the moth in the alcove you didn't take from to begin with.

Now; the issue of getting back up without switching from Lumos to Depulso and losing the moth. What you need to do is place the moth you're holding on the moth podium on the left side of the room, then Depulso whilst on the rising platform to get back up to the moth door. From there, Depulso the mechanism a number of times and you'll see that that podium rises as the mechanism turns, meaning you can cast Lumos from right next to the moth door to grab the moth and place it in, allowing you to progress through.

In the room down the steps, you'll find the witches resting place. At the coffin, you'll see a dead Ashwinder who you can search for a Signet Ring, though the Helmet has already been taken.

Once you're ready to leave, go behind the coffin and interact with the wall to exit out to the North Hogwarts Region.

As you arrive outside, speak to Lodgok who will tell you there is an encampment nearby, and it is your best chance to get the helmet before Rookwood escapes.

Recover the Helmet from the Thieves

Once you arrive at the camp, clear out the enemies and go to the open tent on the right side of the camp.

Loot the large chest there to find the Helmet of Urtkot, and you'll be ambushed by some Ashwinder Assassins. Deal with them, and the Inferi they summon, then return to Lodgok with the helmet.

Speaking to Lodgok, he will thank you and says he'll take it to Ranrok to distract him from his search - potentially slipping up that he knows more than he is letting on. As the conversation ends, the quest concludes.

✧ The Elf, The Nab-Sack, and the Loom

Meet with Deek

Return to the Room of Requirement, and a short cutscene will play where Deek finds the Nab-Sack, and once you regain control, speak to Deek next to the Room of Requirements Floo Flame location and select "I'm ready to learn how to use the nab-sack."

Rescue a Puffskein

You'll teleport outside of Hogwarts, and will need to follow Deek to the nearby Puffskein den. You will then receive the nab-sack item, and will need to assign it in a Spell Slot. Equip it, approach a Puffskein and use it while aiming at the Puffskein. Wait for the golden circle to complete, then hit the button prompt to rescue the animal.

Once successful, speak to Deek once again, and he suggests capturing a Jobberknoll next, providing you a nearby location. Head there and speak with Deek on the opposite side of the bridge to the Jobberknoll tree. Deek recommends that Levioso can slow down flying beasts, making them easier to capture.

Rescue a Jobberknoll

Cross over the bridge, and slowly approach the Jobberknoll tree. Select your target and cast Levioso on a Jobberknoll, and quickly equip and use the nab-sack, this time having to hit two button prompts to successfully capture the beast.

With that done, return to Deek who will speak of Mooncalves also being nearby. Head to Deek's new location nearby and speak to him. He will explain that Mooncalves only come out at night, meaning you'll need to wait. Open up your map and hit the Wait button to turn day to night.

Rescue a Mooncalf

Now you have waited till night time; approach a Mooncalf, cast Levioso to make the capture easier, then use the nab-sack, hitting 3 button prompts in order to successfully nab the Mooncalf. Return to Deek a final time, and he will state there are other creatures in the world, but first you'll need to return to the Room of Requirement to settle those you have already rescued.

Conjuring the Vivarium

As you travel back to the Room of Requirement, in a cutscene you'll conjure up the Vivarium, a greenhouse within the Room of Requirement that acts as a free-range home for your rescued beasts.

Head inside once back in control, and open up the Beast Management Menu with the on-screen prompts. Select the Puffskein, Mooncalf, and Jobberknoll, and they'll all be released into the Vivarium.

Deek informs you that a beast shop in Hogsmeade can help care for beasts you don't have space for, and that if you care for the beasts in your Vivarium, they will occasionally deposit a related material for you to collect; Mooncalf Fur for example.

Collect Feathers and Fur

Now equip your Brush and Feed, then one by one provide both for the Puffskein, Mooncalf, and Jobberknoll. Brushed and fed, each animal will provide their respective materials; Puffskein Fur, Mooncalf Fur, and Jobberknoll Feathers. After collecting those three things, speak to Deek once again. He will provide you with a Spellcraft for an Enchanted Loom that lets you alter pieces of gear.

Exit the Vivarium and equip your Conjuration spell. Head to the Utility tab and select the Enchanted Loom. Place it down in your Room of Requirement, and interact with it. You'll need to add a trait to a piece of clothing, so select a piece of gear in your inventory with a blacked out circle, then choose from the list of Traits you have collected thus far.

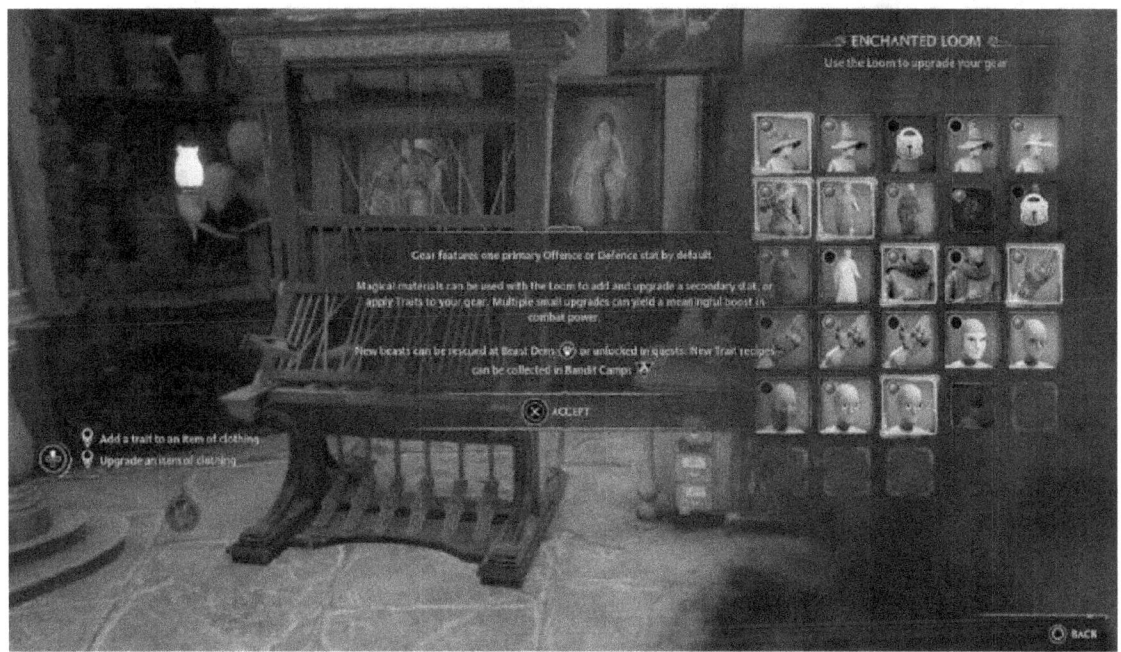

Then select a piece of gear you'd like to strengthen (likely a piece you currently have equipped), and go to upgrade it once. With those two things done, speak to Deek again, after which the quest will conclude.

✧ In the Shadow of the Estate

Meet Sebastian in Feldcroft

Head to the southwest of the map to the small hamlet of Feldcroft, and meet with Sebastian in the wooden tower there. He'll ask you to follow him to meet with his sister Anne, and his uncle in the small cottage nearby.

In a cutscene, tensions rise between Sebastian and his uncle, whilst Anne suffers. You'll need to speak to Solomon Sallow and Anne Sallow separately. Solomon is inspecting a cart outside, and Anne is sat at a table inside.

The Goblin Assault

With that done, speak to Sebastian just outside the cottage, and he will ask you to follow him to where Anne was afflicted with her condition. As you leave the hamlet, you'll be interrupted by a large Goblin ambush, which you'll need to defeat in order to progress. Partway through the battle, a Loyalist Commander arrives.

With all the Goblins defeated, speak to Sebastian who will explain that the area you are in now is where Anne was cursed. It seems the Goblins were trying to hide something, and you both agree to search the surrounding area, as well as Rookwood Castle.

Inspecting the Area

Use Revelio to inspect a variety of items around the area, progressing up to the destroyed house and investigating the portrait inside. Then talk to Sebastian to reveal to him that this house is the one you saw in the Pensieve.

After the conversation concludes, hit the stack of debris in the wall with Depulso and crouch through. Head down the stairway to your right, and begin picking up Isidora's diary pages.

Break down the bookcase on the right to progress further into the cellar, and you'll see a mirror in the corner. After discussing that you can see the Undercroft, speak to Sebastian once more. You discuss your Ancient Magic abilities, and once your back in control, head through the mirror into the Undercroft.

Inside the Undercroft

Once inside the Undercroft, you'll see a hidden compartment reveal itself on one of the walls. Inspect the Rune Diagram page inside, then show it to Sebastian. After considering what the symbols could mean, the conversation and quest concludes.

✧ Astronomy Class

Attending Astronomy Class

Head to the Astronomy Tower and begin the class by interacting with the golden circle just outside the Astronomy classroom. Here you formally meet Professor Shah and Amit Thakkar as you stargaze.

Follow the on-screen prompts to focus your telescope, as Professor Shah encourages you to stargaze often outside of class.

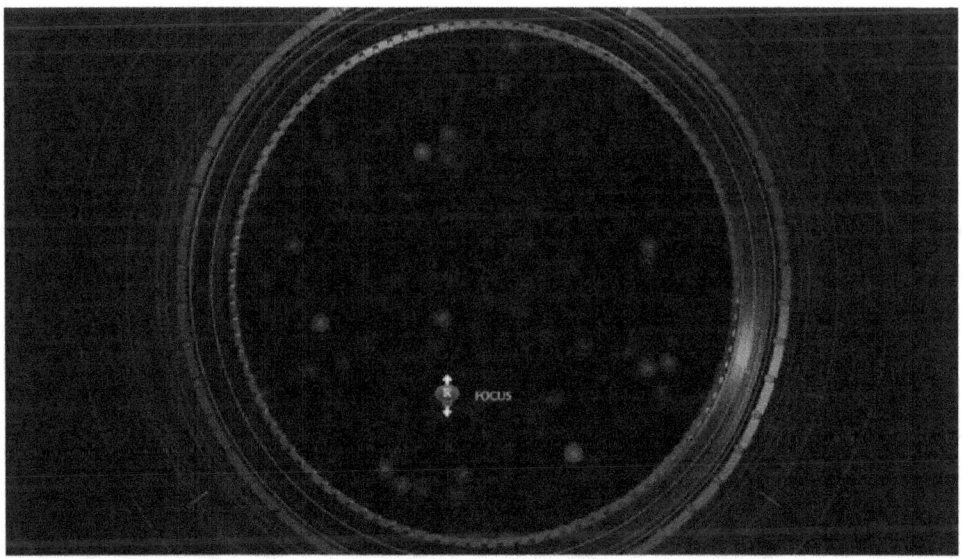

After the lesson concludes, Amit promises to lend you his spare telescope, which you must then retrieve by going down the small set of stairs on the right side of the observation deck, and taking the telescope from the table.

Head back up and then down the main stairs one floor to speak with Amit again.

Amit informs you about the Astrology Tables which can be used to discover hidden constellations. You will teleport outside of Hogwarts, where you'll need to follow Amit up to the ruins of the wall. Take the lead when instructed and break the boxes to the door once inside, continue across and into the next room, using Incendio or Confringo to burn the cobweb blocking your path, then interact with the Astrology table to begin the minigame.

Astrology Table Minigame

Once you've interacted with the Astrology Table, you'll want to zoom in/out, rotate and move the telescope so that the outline of the constellation lines up with the stars in the constellation itself.

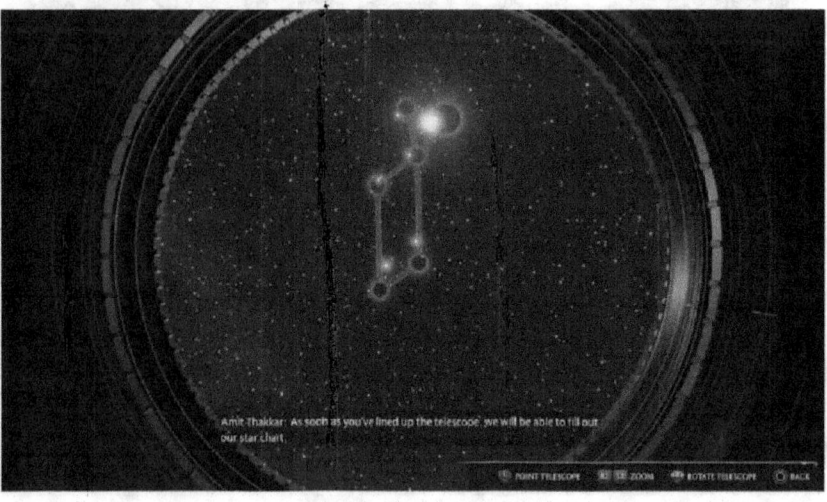

Upon doing so, it will reveal the Lyra constellation, and Amit will congratulate you. After he says you can keep the telescope and use it on the other Astrology Tables in the world, the quest will conclude.

✧ The High Keep

Meet Natty near Falbarton Castle

Heading to just outside of Falbarton Castle, speak to Natty and she will explain that her investigation into Harlow has borne fruit. There is supposedly a letter within Falbarton Castle that proves the collusion between Rookwood and Harlow, which could help to lock them both up.

Entering the Castle

After the conversation is over, head forward and to the right to climb up the battlements. Hit the mechanism on the left with Depulso, then Wingardium Leviosa the crate inside over to the right and then Levioso it, climb on top and climb again to reach the upper rampart.

Hit the wooden boards to free a crawlspace, but before you crawl through, head around the right and use Accio through the window on the crate in the room. Return to the crawlspace and head through.

Once inside, hit the gate mechanism multiple times with Depulso, and then quickly use Accio to pull on the ring on the wall to block the gate from fully closing.

With Natty joining you, walk with her up to the gate to witness a cutscene where Highwing the hippogriff breaks out as dark wizards hit it with spells.

Making Your Way to the Roof

Climb up the battlements to the right, then cast Depulso through the broken window on the crate at the end of the walkway.

Drop down to the right, then Wingardium Leviosa the crate into the gap between the broken stairs on the far side, then Levioso it to create a climbable platform between the two. Once you've reached the top, you'll encounter a number of Poachers that you can deal with as you wish.

Use Alohomora to unlock the gate on the left, and continue up around the path. Continue around and some Poacher Rangers will ambush you. Deal with them, and again use Alohomora to unlock the gate. Inside the tower, there are a couple of Puffskeins and a Niffler you can rescue if you wish, but otherwise continue all the way up the staircase to find Highwing and another Hippogriff. You climb aboard Highwing, and Natsai is picked up in Highwing's claws before being dropped on top of the other Hippogriff, just escaping the poachers firing at you from behind.

Flying High(wing)

Here you gain control of a flying mount for the first time, as you glide with Highwing through the air, past Hogwarts before eventually landing down, celebrating your successful escape. Natty explains that Rookwood is looking for a Phoenix, and will be in touch soon with further information about the letter.

✧ Back on the Path

Return to the Map Chamber

Speak with Professor Fig and inform him of your plans with Lodgok, and Charles Rookwood informs you he saw some Goblin activity in Rookwood Castle, which also happens to be the location of the second Trial.

You inform Charles that his descendant, Victor Rookwood, is in league with Ranrok. Charles asks that you find his portrait within the castle and seek out a source of power before it falls into the wrong hands.

✧ Charles Rookwood's Trial

Meet Fig at Rookwood Castle

Head to Rookwood Castle in the extreme southwest of the map, and speak with Fig near the entrance. In order to gain entrance, you can approach stealthily or go loud, but regardless of your choice, you will be ambushed by a Loyalist Ranger as you curve around the left side of the castle.

After defeating the couple of enemies that spawn in, use Accio/Wingardium Leviosa on the crate and move it near the crumbled part of the castle walls. Then cast Levioso, climb onto it, and enter the castle proper.

You'll then see a cutscene of Ranrok and Rookwood deliberating their roles in their tenuous partnership, and you learn that Ranrok knows of the Keepers, and Rookwood is seeking out Ancient Magic stores.

As you regain control, you'll be ambushed by a very large force of both Ashwinder enemies (including an Executioner), and Goblin Loyalists. Defeat them all, then proceed up the steps on the other side of the courtyard and loop around the ramparts and through the open door.

Inside Rookwood Castle

Drop down onto the wooden platform, which will partly collapse, then drop further down onto the ground and you'll see a locked door with the Ancient Magic symbol. To your right will be two metal symbols on the wall nearby one another, with the third being to your left behind some breakable wooden boards.

Hit all three symbols with your Basic Cast in quick succession to open the locked door, letting you proceed downwards to the bottom of the steps, where you'll need to interact to enter the Rookwood Cellar.

The Rookwood Cellar

Once in the cellar, proceed left through the gaping hole in the wall to find a depleted Ancient Magic reserve. Head through the opening on the right and into the next room, where you'll find Charles Rookwood's portrait.

Speak to the portrait and Charles will urge you to begin the second Trial. Once back in control, interact with the Ancient Magic spot just to the right to conjure the magical archway, and you can enter through the door at the end of the corridor into the second trial.

The Second Trial

As you enter the Trial, head up the stairs on the right and interact with the Ancient Magic spot there. Head back down the stairs and use Accio to pull a box out from the left corner of the room (underneath the exit you're trying to reach). Place that in the middle of the gap between the Ancient Magic spot and the other side.

Then, head to the other side of the archway portal in the center of the room, and you can Accio another box through the portal to your side.

Now, Wingardium Leviosa this second box on top of the first, and then pass through the portal to turn the boxes into a solid pillar that you can climb on, and then climb up again to reach the next area.

As you enter the next room, you can hit a distant symbol on the wall in an alcove to the left to create an additional section of bridge, letting you reach a chest. Otherwise, continue on and interact with the Ancient Magic spot just to the right of the archway. You'll see the archway activate, and a symbol/mechanism appear on the wall to the right.

You can hit this mechanism with Basic Cast to rotate the archway 180 degrees, which you'll need to do to Accio the pillar (which turns into a box on the other side), through the portal and out of the way, letting you spin the archway again to now proceed through the open walkway.

In the next area, you'll see a Pensieve Protector on a circular platform, so give it everything you've got to defeat it, though be careful as both Pensieve Sentries and Sentinels will drop into the fight shortly after it begins. Take out these smaller enemies first as their ranged attacks can catch you off guard, and you can get swarmed easily. With those out of the way, deal with the big one. Be careful throughout the duration of this fight, as the circular arena isn't very wide, and if you get caught in an aerial shockwave attack (denoted by the blue pulsing flame circles on the floor), then you can easily be knocked off to your death.

Moving forward, interact with the Ancient Magic spot in front of the archway ahead of you, and hit the symbol through the portal with a Basic Cast.

Head through the portal, now to your right, then hit the symbol on the wall again with a Basic Cast to move the portal out of your way, revealing a now intact path on the opposite side.

As you cross over the left side, interact with the Ancient Magic spot to the right which will activate the archway portal in the center of the room, as well as reveal a pillar to the left of you. Drop down into the center of the room, and immediately sprint through the side of the portal facing you. There are 2 Pensieve Protectors in this room, that are invisible/impervious to damage unless you are on the correct side of the portal.

Deal with the one you've revealed by going through the blue side of the portal (whilst avoiding attacks from the other invisible one), then pass through the red side of the portal and deal with the second Protector.

With the invisible threats vanquished, ensure you are on the blue side of the portal (you'll know by the color of the wisps around the edges of your screen), then Wingardium Leviosa the box through the portal to the red side.

Return the box to where it was on the ledge, then pass back through the portal to the blue side, where you'll now see that the pillar AND the ledge you're trying to reach will both be tangible at the same time

Climb the pillar and then onto the platform, and hit the symbol beneath you to the right with a Basic Cast to rotate the archway/platform once more, letting you jump across the gap and exit the room.

In the next room down the stairs and to your left, a couple of Pensieve Sentries will activate. Deal with them, then head into the room ahead of you, up the stairs, and across through the door to arrive at another circular arena.

As you reach the center, the archway portal will activate, and you'll have another situation with two Pensieve Protectors, one of which will be invisible till you travel to the other side of the portal. The catch here is that there are also Pensieve Sentries and Sentinels to contend with, some of which will also be invisible. Focus on one side of the portals enemies first, clear them out (evade attacks from all invisible enemies), then quickly transition through the portal and clear out the remaining enemies now that you can see/affect them.

After all enemies are defeated, a giant Pensieve Guardian will emerge from the ground. Dodge its stomps hit the orbs it conjures with the corresponding color magic before it is launched at you, and in the second phase be sure to parry at the correct time when it swings its flail upwards and down, shockwaving the entire arena.

Once the Guardian is down, cross over the bridge that forms to enter the chamber with the large statue and Pensieve. Approach the Pensieve and interact with it to view one of Charles Rookwood's memories.

The memory in question is of Isidora welcoming the other professors into her home, where she introduces her father to them. Isidora claims to be able to take away pain using magic, to which she demonstrates on her father. The other professors chastise her, though it appears to have worked.

With the memory over, use the Enchanted Stone archway to exit the Trial and return to the Map Chamber Once there, speak with Rackham's portrait. Both he and Rookwood will commend you for completing the Trial, and they introduce you to Niamh Fitzgerald, the former Hogwarts headmistress.

Whilst she prepares the third Trial, you are encouraged to hone your magical abilities further. Speak with Fig briefly to end the scene, with the season outside transitioning to winter.

✧ Fire and Vice

Meet with Poppy

Head to the north of the Forbidden Forest and meet with Poppy Sweeting. She asks about Highwing, and also explains about Horntail Hall, a poachers' den where they exploit creatures for their own entertainment.

Find Horntail Hall

Follow Poppy a little up the path to encounter a group of Centaurs, chastising you for not staying in Hogwarts and accusing you of collusion with the poachers. After reassuring them you aren't with the poachers, they bid you farewell with a word of caution.

Continue down the path with Poppy, watching a Dugbog drag a deer into the lake; unusual behavior, as Poppy points out. A little further ahead, you'll need to investigate the poacher camp. Check the cage, the pelts, the fire, and the goblin metal behind the tent.

With that done, return to Poppy. After a brief conversation on what you found, proceed forward and down the hill. Ahead are a number of Poachers, which you can choose to approach stealthily with Disillusionment, or outright aggression. If the bridge breaks, use Reparo to put it back together. Across the bridge, enter the tent to trigger a cutscene where the truth of Hornstail Hall is revealed.

Inside Horntail Hall

Horntail Hall is a dragon fighting ring. You and Poppy agree to discreetly look around, though you can choose to be loud if you wish. Whatever your approach, deal with the poacher in front of you, then head into the room to your right, and then down the steps. Another poacher is down here, with two Loyalist Goblins. Deal with/evade them, and climb the ladder on the far side of the room.

Continue ahead, but before turning right you'll trigger a short cutscene where you see a third dragon in chains, being tormented by Poachers.

Go down to the right now, and open the door. On the table in the center of the room, use Alohomora to unlock the cage containing a Dragon Egg.

With the Dragon Egg now in your possession, turn around and head down the steps into the arena with the chained dragon.

Here you'll be fighting a variety of enemies, mainly Poachers and Goblin Loyalists (and their various iterations). The Duelling Feats are to keep two enemies airborne simultaneously for 5 seconds, disarm an enemy (with Expelliarmus), and to disrupt the Animagus' Reducto cast with Depulso.

With the two waves of enemies defeated, use Accio to pull on the chainlink at the base of the restrained dragon, freeing it. During a cutscene, you are surrounded by enemies by the dragon burns them all to a crisp, as you create a hole in the ceiling of the tent for the dragon to escape through.

As you and Poppy escape, she exclaims it was a Hebridean Black, and the Dragon Egg likely belongs to her. After a brief conversation, the quest concludes.

✧ In the Shadow of the Mine

Meet Sebastian by the Overlook

Head to the Overlook just northeast of Upper Hogsfield and speak with Sebastian. He will explain that this is the area painted on the triptych earlier, and that there is an abandoned mine nearby that is a hub for Ranrok's Loyalists. You can choose to approach the mine stealthily with Disillusionment or go in full force.

Go down the hill ahead of you and you'll need to pass through two separate encampments of Loyalists. Deal with both as you wish; loud and proud, stealthily eliminate enemies, or avoid them all together, to reach the entrance to the Overlook Mine.

Inside the Overlook Mine

Once inside the mine, follow the path round to the left after dealing with the two Loyalists by the edge, jumping the gaps and clambering up before sliding down the slope ahead.

You'll land in a room with three Loyalists, who you'll need to eliminate before progressing.

After the room is cleared, use Wingardium Leviosa to lift the crate in the room against the wall on the west side of the room, letting you climb up to the higher area.

Go down the path to the left to slide down a second slope.

At the bottom, climb up the ledge on the left and cast Reparo on the broken bridge, allowing you to cross. On the other side, you'll be ambushed by 3 Thornback Scurriors. Defeat them and continue through the passageway and crouch under the gap to the left to be ambushed by another spider.

Continue through the next crawlspace and follow the path before you'll drop down into a large combat arena.

Here you'll face a large number of Loyalists and then Thornback Spiders, including two Matriarchs. Deal with them all as you see fit.

With the room clear of enemies, use Confringo/Incendio to clear the various spiderwebs and you'll note three rune symbols. One directly next to the locked rune door, one just above it to the left, and one further off to the right. Hit all three in quick succession with Basic Cast to unlock the door.

With the door open, head down the stairs and explore the small chamber. In a chest is the second triptych piece, and a journal entry from Isidora is on the table in the center of the room.

Speak to Sebastian after obtaining these to confirm plans to return to the triptych to see what you can reveal. You can then use the enchanted stone reflection within this chamber to teleport directly to the Undercroft.

Returning to the Undercroft

Back in the Undercroft, walk forwards and place the second piece you've just recovered into the triptych on the wall, then speak to Sebastian. Sebastian recognizes the location, but warns there is more trouble waiting should you head there. He chastises you when you discuss your plan with Lodgok, and tells you to leave in a dismissive manner, concluding the quest.

✧ It's All Gobbledegook

Speak with Amit

Head to the top of the Astronomy Tower and speak with Amit as he is one of the few you know that can speak Gobbledegook. You ask for his help and he excitedly agrees, saying he will meet you and Lodgok outside the mine.

Meet Lodgok Near the Mine

Go far south, just below Keenbridge, and meet with Lodgok near the Mine's Eye Floo Flame location. Speak to Lodgok who claims the Helmet plan didn't work, and Bragbor, an ancestor of Ranrok's, had repositories of Ancient Magic built across the world, and Ranrok is seeking them out.

Amit arrives shortly thereafter, and Lodgok explains you'll need to read Goblin plans whilst evading the eye above the enchanted door and the roaming Loyalists. Lodgok explains he can't come with you for fear of being reported to Ranrok.

Depart and approach the nearby mine entrance, casting Disillusionment to evade detection by the eye above the door. Once you open the door, you can then enter The Mine's Eye.

Inside the Mine's Eye

Now inside the mine, head down the path and hit the fireplace to your left with Confringo/Incendio to open up the elevator nearby which you can then activate the mechanism inside of to travel down deeper into the mine.

Head down the passageway to the right and use Accio to drag the platform over.

Walk aboard, then drag yourself back across to the other side. Up ahead you'll find some Loyalists to defeat, and another eye door which you'll need to cast Disillusionment to get through.

Follow the tunnel round and inspect the scroll on the table in the main room you emerge out into.

Another scroll is on the table on the first floor, to the right.

You'll find another later in the area on a barrel just to the left of a slope, leading into a combat area with multiple Loyalists, including a Commander.

Clear them out, then check the right-hand side of the arena for another schematic near the open eye door.

On the left side of the arena, light the furnace with Confringo/Incendio, then pull the mechanism up to the right with Accio to open the waterway tunnel behind it.

Proceed through the cave and crawl through the narrow space to reach another combat arena. Clear out the Loyalist enemies, then cast Disillusionment to sneak up to and through the eye door.

Go up the steps to the left in the room ahead, and inspect the scroll on the table for a schematic of a gigantic drill.

To exit the mine, light the furnace in the same room as this schematic, then turn left immediately after exiting the eye door to pull a mechanism with Accio, activating the lift.

Riding the lift up and backtracking out of the mine to Hogwarts Valley, speak to Amit once outside who will bid you farewell, letting you speak to Lodgok nearby. You inform him of the Goblin's drill plans, and he explains that he had an extensive history with Miriam, Professor Fig's wife, and due to her kindness, he decided that Ranrok's approach is not the way forward. Lodgok warns that if Ranrok discovers the repositories, a great war is bound to ensue.

✧ Speak to Niamh Fitzgerald

Head to the Map Chamber and speak with Niamh Fitzgerald's portrait to learn that the third Trial is located in the Headmaster's Office, and you'll need to access it whilst Headmaster Black is away.

To learn of a way in, you decide to head to Professor Fig. Speak to Professor Fig in his classroom to begin The Polyjuice Plot quest.

✧ The Polyjuice Plot

Take the Polyjuice Potion

Speaking to Professor Fig and catching him up on what you've learned about the drills Ranrok is creating, and that Lodgok knew Miriam, you also inform him of the location of the third Trial. Fig reveals he already had a Polyjuice Potion for a Professor Black disguise prepped ahead of time, for any eventuality, and you take it.

Turning into Professor Black, Fig tells you to seek out Black's House Elf - Scrope, to learn the secret password to gain entry to the Headmaster's Office. Your best bet is to speak to Madam Kogawa who has been pestering Scrope to help her reinstate Quidditch.

Speak to Madam Kogawa

As you depart Professor Fig's classroom, Professor Sharp will intercept you. After a brief conversation, make your way down to the bridge connecting the Great Hall and the Library Annex. On the bridge, you'll speak to Madam Kogawa, who will inform you that Scrope is hiding in the Great Hall.

Continue on towards the Great Hall, and as you reach the corridor prior to entering the Great Hall, Professor Weasley will interrupt you with concerns about Fig and "the fifth-year's" escapades. Assuring her there is nothing to worry about, proceed into the Great Hall, where Scrope is standing at the lectern by the giant Christmas tree.

Get the Password from Scrope - Black Family Motto

Speaking with Scrope, he explains that "you" swore never to reveal the password to anyone ever, but after some aggressive convincing he explains it is the Black family motto. Choose the "purity of blood" option, and Scrope explains that it is actually "Always Pure" in French - Toujours Pur.

With the password obtained, head up the steps on the left side of the Great Hall, and interact with the folding screens in the corner of the room to change back to your usual self.

✧ Niamh Fitzgerald's Trial

Enter the Headmaster's Office

With the password obtained, head to the Trophy Room and unlock the double gates blocking the stairwell (if you haven't yet already). Continue all the way up and you'll eventually arrive at the statue of a Phoenix on the left hand side.

Whisper the password there and the spiral staircase to the Headmaster's Office will appear. Climb up and head forward towards the desk to trigger a cutscene where you speak with Niamh's portrait. She tells you to read the book at the back of the room, up the steps.

Inside the Book

As you interact with the book, you will be magically sucked inside. Witnessing the fable of the Deathly Hallows from your own unique perspective, head towards the village to see a gigantic hooded skeleton emerge, and summon minions that will kill you on sight.

This is a lengthy stealth section that is far easier than it seems. With Disillusionment being auto-cast, progress through the village whilst taking cover arriving at each new set of enemies. A window will open where all are facing/moving away and that is your time to move.

After going underneath a bridge, you'll need to climb a ladder to the left as deadly dogs spawn in ahead to the right.

Walk over to the other side of the roof and drop down, then continue to evade enemies as you have before. As you move through a number of dillapidated buildings, you'll eventually drop down with a courtyard ahead of you, in the middle of which is the legendary Invisibility Cloak.

Interact with it to equip it, allowing you to continue forward and follow the path, now being able to walk directly past enemies without being seen. Follow the linear path and eventually the scene will reset to white, this time with a Mysterious (familiar looking) Wand in front of you.

Using the Wand

Now back in the storybook segment, cast Bombarda on the debris blocking your path, then press on. As you approach the castle entrance, 5 Death's Shadow enemies will appear, which you'll need to defeat however you see fit.

As you enter the castle after clearing those enemies, another wave of Death's Shadows will descend, alongside Shadowy Mongrels. Beat them all and then head inside the castle proper. Up the stairs, another arena awaits.

There will be two waves here. The first consists of Death's Shadows and a total of two Death's Trolls, one after the other. In the second wave, you will have Death's Shadows, Death's Dark Mongrels, and two Death's Trolls all at once.

After all the enemies are down, head to the left side of the arena, down the steps and onto the podium to return to the white space, with another artefact in front of you.

The Mysterious Stone

Picking up the Mysterious Stone ahead of you, proceed through the cemetery you find yourself in. As you reach the top of the cemetery, the scene will dissipate again except for a body resting on a stone tablet.

Interact with the body to resurrect it, revealing it to be Niamh Fitzgerald herself. Follow her down the path towards the giant statue, again with a Pensieve beneath it, which you can interact with to view Niamh's memories.

The memory pertains to Niamh confronting Isidora after learning that she hasn't stopped her experimentation with magic surrounding emotions and the removal of pain, which Isidora doesn't take well.

As the memory ends, use the enchanted stone archway to return to the Map Chamber, and you'll be introduced to the fourth keeper, San Bakar.

Discussing Isidora inhaling negative emotions in the memory, San Bakar is hesitant to reveal the location of his Pensieve before consulting with the other Keepers. You and Fig decide to keep an even keener eye on Ranrok in the meantime.

✧ In the Shadow of the Mountain

Meet Sebastian Along the Coast

Meet Sebastian on the northwest coast of Marunweem Lake, and he will explain that Ranrok's Loyalists have discovered another cavern nearby, and you both agree that the next piece of the triptych is likely in there.

Head up the hillside along the winding roads towards the entrance to a Goblin camp, that Sebastian will promptly charge into with no hesitation. Defeat all the enemies in the area as you see fit to have a brief conversation with Sebastian where you chastise him for his brashness, but he doesn't listen.

Continue Up the Mountain

Continue up the hill, climbing where necessary, and defeat the couple of Goblins you encounter on the way. You'll shortly thereafter arrive at another small Goblin camp, which you can now approach steathily if you wish. Deal with all the enemies in the area and another conversation with Sebastian ensues. You eventually convince him to calm down and listen to reason, and can then proceed up the steps to enter the Tower Tunnel.

Inside Tower Tunnel

Once inside the tunnel, proceed along the path and you'll be ambushed by various Thornback Spiders, including a Matriarch. Deal with them all as you see fit, and at the fork in the path just ahead (with rocky debris to the left), continue down the path to the right.

You'll quickly encounter a few more spiders, before coming upon a staircase.

Head up the stairs, clearing the cobwebs with Confringo/Incendio, then Depulso the debris blocking your path. You'll now arrive in a spacious room where dozens of spiders will descend from the ceiling, including two Matriarchs. Eliminate them all, then take note of the Rune Door just up the steps. Two of the three symbols you'll need to hit with your Basic Cast are on either side of the door (the right one is covered with cobwebs you'll need to clear first), with the third directly behind you, embedded in the wall (if you're facing the door head on).

Hit all three in quick succession with your Basic Cast to unlock the door, allowing you to move on. In the corridor ahead, you'll find another of Isidora's journal entries in a crate against the wall. At the end of the corridor you'll find a giant hole in the wall, which you can slide down to initiate a boss encounter of sorts with a Mountain Troll and a few spiders.

Defeat them all, then cast Reparo on the large amount of debris on the right side of the room to fix the walkway.

Head through into the next room via the now clear walkway and Accio/Wingardium Leviosa the crate over to the destroyed staircase in the room. Cast Levioso once it is in place, then climb onto it, and up onto the stairs to progress.

Continue up the stairs, then loop around the right side of the room with the sets of armor, and up a small set of steps to enter another large room with sunlight pouring in. Up the slope of debris on the right side of the room, cast Depulso to clear the path, and jump over the gap to reach the next room.

Around the corner to the left is another Rune Door. One of the symbols is tucked behind some debris immediately left of the door, so cast Depulso to clear that debris out. The other two symbols are behind you, hanging over the dropoff that you passed to reach the door.

Hit all three in quick succession with your Basic Cast to unlock the Rune Door. In the room beyond, you'll find a base that Isidora set up for herself. Climb up to the second floor using the perches just to the right, and grab the final triptych piece off the table up there.

You can then use the enchanted stone wall nearby to travel directly to the Undercroft. Once back in the Undercroft, place the final piece of the triptych in the wall to reveal a secret Pensieve of Isidora's.

It is the same memory of her removing the pain from her father that you saw from Rackham's perspective with the addition of Bragbor the Goblin arriving shortly after the others leave to help Isidora safely store the pain and magic she removed until it can be used later.

Exiting the memory, Sebastian is convinced that this is how you and he can help save his sister, Anne, and though you aren't sure of your abilities or the Keeper's willingness to teach you the methods, you agree to keep an open mind and at least entertain the idea for the time being.

Enter the Mine

Head to the north end of Marunweem Lake to find the Coastal Mine. This is a large encampment filled with enemies, so be prepared for an arduous fight. As you approach, even an Armored Troll will emerge from near the mine entrance which you'll need to take down.

After clearing the area of enemies, head up the main pathway into the cliff to find a small doorway to the right, allowing you to enter the Coastal Mine proper.

Inside the Mine

Head down the path and deal with the two Goblins you'll find before accessing the tram system tucked to the right.

After a short ride, you'll get off, and will need to head up the small staircase ahead and use Confringo/Incendio to light the fireplace beneath the mechanism. This will allow you to unlock an optional door further into the mine.

Head up the steps near the fireplace you've lit, and use Accio to pull the platform towards you. Get onboard, and then Accio again to pull you and the platform back across. In this new area, which will have a few Goblin enemies to deal with, you will have a number of branching pathways to take. Left, straight ahead, right, or looping back on yourself (just to the left).

Optional Paths

Going right has the option of a small chest to loot alongside some Moonstone, and going left allows you to reach a Collection chest and a few others by dropping through a hole in the ground. If you head straight forward into the covered area, you'll need to light another furnace with Confringo/Incendio which will raise an elevator just down the steps. Board it and activate it to arrive in a fight with multiple Goblin enemies, a Fortified Troll, and the Infamous Foe, Grodbik.

Continuing Into The Mine

Whether you go down any of the optional paths, or just want to continue into the mine, head down the stairs in the area you've just defeated the first few enemies in, and you'll need to Accio the platform at the bottom up to you. Get aboard, then Accio yourself down with the platform.

In this area, more Goblin enemies await, and once they're dealt with you can either board another tram or enter a room to the right that has a fireplace you can light. Riding the tram, you can reach the very first door you unlocked via fireplace earlier in the mine, which holds a Collection Chest inside.

Return and light the fire under the mechanism in the room (now to your left), and ride the lift up that it activates. At the top, in the room ahead, you'll find Ranrok speaking to his Loyalists about the drill's success. As the cutscene ends, defeat all enemies in the area, including an Armored Troll, and then destroy the 5 pillars around the room to bring the ceiling down on the drill.

A cutscene will ensue where Ranrok confronts you, before Rookwood brings in Lodgok. It's revealed that Lodgok is Ranrok's younger brother, and that Lodgok has a book with the location of the final repository, which he gives to Ranrok to stop the fighting. This plan obviously does not work, with Ranrok striking Lodgok with Dark Magic, killing him, as Rookwood unsuccessfully tries to kill Ranrok with Avada Kedavra, whilst you escape on a nearby tram as the ceiling truly starts to collapse.

As you regain control, you can interact with the gate ahead to open it, and turn left up the slope to exit the mine.

✧ San Bakar's Trial

Head to the Map Chamber

As you explain that Ranrok has Bragbor's journal with the last repository location to the Keepers and Professor Fig, they explain that the final Trial requires a lot of skill, and nuanced handling of beasts.

Cragcroft Shore

Travel to the Cragcroft Shore Floo Flame location north of the body of water to the right of the Manor Cape region. Meeting Professor Fig there, you'll need to cast Confringo on the large amount of foliage covering the stone carving ahead of you.

Doing so reveals a carving of a Graphorn, which explains the need for handling of beasts. As you set out to encounter and subdue the "Lord of the Shore", Professor Fig departs for the Map Chamber, as you must face this alone.

The Lord of the Shore

Head to the extreme southwest of the map, in the Clagmar Coast region, and you'll happen upon a large whale skeleton. Proceed into the den ahead to trigger a cutscene where the Graphorn known as the Lord of the Shore emerges. This begins a two-phase boss battle.

You can use your entire arsenal of spells, Ancient Magic, and Ancient Magic Throws to take down the Lord of the Shore. Something to be aware of during this fight is that the Graphorn's attacks are very fast, and often have little telegraphing, meaning you'll need to be ready to dodge at a moments notice. The game's lock-on feature is a lifesaver in this encounter, so be sure to make use of it as the Graphorn charges and leaps toward/around/behind you.

At 50% health, the Graphorn supercharges and becomes even more aggressive, dealing heavier blows that come at a much quicker rate. It will also pursue you more intensely, giving you even less respite than in the first phase. After you eventually whittle him down, a scene will begin where the Graphorn starts charging You are given two options, "Kneel" or "Attack". Choose "Kneel" to bow before the creature, stopping it in it's tracks, allowing you to ride the Graphorn as it becomes your ally, and a permanent mount option from this point forward.

San Bakar's Pensieve

Now riding the Graphorn, run with it back to the Pensieve Chamber you visited with Fig at the beginning of the quest, using the Graphorn's charging ability to ram through any blockades that the Poachers have placed down on your path.

As you arrive at the Pensieve Chamber, stand with your Graphorn on the circular platform on the floor, gradually lighting it up, triggering a hidden portal doorway to reveal itself beneath the stone carving.

Dismount the Graphorn and head inside across the chamber to speak to San Bakars portrait, where he finally relents that you are worthy of this knowledge he's been reluctant to provide up until this point.

Continue down the stairs into the room beyond to view San Bakar's memories via the Pensieve. San Bakar visits the Morganach home to find Isidora's father devoid of all emotion, which he then relates to Rackham who has heard Isidora is even using her magic on her own students, which causes Rackham and the other Professors to confront Isidora near one of her repositories. After an argument, and a brief wand battle, San Bakar casts the killing curse, Avada Kedavra, killing Isidora.

As the memories end and you exit the Pensieve, turn around and use the enchanted stone passageway to return to the Map Chamber. As you arrive, approach Rackham's portrait to speak with the Keepers. It's revealed the repository in the memory, the final repository, is in the caverns beneath Hogwarts.

Having finally proven yourself, Rackham explains you will first need to create a wand using the Artefacts you found above each Keepers Pensieve. Professor Fig tells you to visit Ollivanders, but first inform Professor Weasley of all that is going on, as you'll need all the help you can get for the coming battle.

✧ Wand Mastery

Meet with Ollivander

Head to Ollivander's in Hogsmeade and speak with Gerbold to ask him to create the Keepers Wand. He does so in a brief cutscene, and as you leave with the wand, Victor Rookwood will confront you in the street outside. He proposes an alliance to ensure the Goblin's don't succeed, but you refuse his request. Before you can do anything else, one of Rookwood's lackeys grabs you and Apparates away.

Battle Victor Rookwood

In the arena you are teleported to, you will face a wave of enemies consisting of various Ashwinder and Poacher enemies. Keep moving, make effective use of Protego/Stupefy, and eliminate them all. Upon doing so, Rookwood will appear and will trigger a Wand Battle, where you'll need to quickly and repeatedly hit the button prompt on screen to cause your magic beam to overpower his.

He will quickly disappear once more, with more Ashwinder and Poacher enemies arriving, though in this second phase Rookwood will also be on the field. The shield charm Rookwood has is not related to any particular Spell type, so the only ways to break it are: Ancient Magic, Ancient Magic Throws, or the Protego/Stupefy combo.

Once broken, deal as much damage to Rookwood as possible, then rinse and repeat until he is defeated, causing a second wand beam battle, with another button prompt you'll have to mash, before wiping him from the face of the earth.

With Rookwood gone, and the Keepers' Wand safe, you must return to the Map Chamber and speak to Professor Fig of your success.

✧ The Final Repository

Meet Fig in the Map Chamber

As you meet and speak with Professor Fig in the Map Chamber, Rackham explains you are finally ready, and must resist the temptation to destroy or harness the magic within the repository. The map in the floor dissipates and descends, revealing a secret staircase and a door. Head down the stairs and through the door to enter the Keepers' Caverns.

The Keepers' Caverns

Upon entering, you must follow what is essentially a long winding linear path through the caves, dealing with enemies as they emerge. You'll eventually enter an arena with two Fortified Trolls alongside regular Goblin Loyalists, all of whom you'll need to defeat.

Once clear, head towards the stone wall to the back of the arena with the glowing white smoke for Fig to blow it up, letting you progress.

After another arena down the path, this time with just Goblin Loyalists, head to the right where another glowing white smoke rock can be blown up by Fig.

In the gigantic cavern beyond, a cutscene will trigger where the various Professors of Hogwarts will Apparate in and assist in the battle.

Back in control, head across the bridge to a small combat arena against some Loyalists, and then a Fighter Troll will break through. Bring them all down, and after a lucky hit from Professor Sharp creates a bridge you can use, you and Fig reach the entrance to the repository.

Two Pensieve Guardians stand at either side of the door, and before they can strike you bring out the Keepers' Wand and they stand down, with the door opening ahead.

Once inside, Fig will ask what you intend to do with the magic inside the repository. You can choose "I intend to keep it contained here." or "I intend to open it.".

NOTE: This dialogue choice will dictate which ending you get, but fear not, there are no real ramifications or consequences, or even any major differences in cutscenes between the "good" and "bad" endings, so simply choose whichever narrative conclusion you'd prefer.

Good Ending

Choosing "I intend to keep it contained here" causes Fig to breathe a sigh of relief, knowing your plan is to keep it contained and not repeat the mistakes of Isidora, and you can then choose whether to keep it a secret "for now" or "forever". Regardless of your choice here, Fig agrees this is the right path. Ranrok arrives shortly thereafter with Miriam's wand, and casts a destructive spell that blows the repository open.

After Fig seemingly dies getting crushed by falling debris, you slip and fall. Down a small slope, the battle against Ranrok begins. Please refer to the Defeat Ranrok section below for a full breakdown of the final boss fight.

❖ After Defeating Ranrok (Good Ending)

After defeating Ranrok, as the dark magic runs rampant, you use the Keepers' Wand to summon a new repository cage to contain the magic with the help of Professor Fig, who shortly thereafter dies from his injuries.

❖ Defeat Ranrok

Ranrok, with the power of the repository unleashed, has taken the form of a shadowy dragon. He is invulnerable to regular spell attacks, so you must keep an eye out for floating magical balls that you'll need to strike with a spell of the corresponding colour.

Only then will he become vulnerable attacks for a brief time. Ranrok's attacks here are very fast, with little time to tell if you can Protego the incoming shot, or if you'll need to dodge.

After you deplete his first segment of health, he will destroy a wall and allow you to progress into the next stage of the fight. Sprint all the way down the path to the second battle stage, where the same process applies of hitting the colored orbs to make Ranrok vulnerable, but now you must hit two orbs before he can be hurt.

Depleting the second segment of health will allow you to slide down into the very bottom of the cavern, in a large circular arena. Though he has 1 large health bar left, this is still divided into two phases. First will be much like the previous two, but three orbs must be hit successfully before you can damage Ranrok. The final phase has Ranrok land down in his dragon form, where he can charge, swipe, and still spit fireballs. Edurus Potions are MVP here if you happen to have any, as you'll likely take some hits whilst you take care of FOUR different orbs before Ranrok can be targeted.

Luckily though in this phase, you only have to hit the four orbs once and he will remain vulnerable for the duration of the fight from that point forward.

Fully depleting his health will have him in the center of the arena, where you must hit a final 3 orbs that all dissipate very quickly, meaning you'll need to switch between spell types quickly or they'll move to a new location in the arena. Hit them all, and Ranrok will be defeated.

Bad Ending

Choosing "I intend to open it" when speaking to Professor Fig, he will be shocked, and ask that you reconsider after all you've learned regarding Isidora's fall into darkness and obsession. If you double down and say the power shouldn't be contained, or that you want it specifically for yourself, then Fig will be in disbelief that this is your choice. Before he can say any more, Ranrok arrives with Miriam's wand and casts a spell that destroys the last repository, freeing all the dark magic and emotion within.

The final boss fight against Ranrok plays out the exact same way, so please refer to the Defeat Ranrok section above.

❖ After Defeating Ranrok (Bad Ending)

Upon defeating Ranrok, as the dark magic runs rampant, you use your wand to summon all of the dark magic towards and into you, channeling it to become very powerful with your eyes flashing red, harnessing it and guiding the remaining dark magic around the room as you will it.

Remembering Professor Fig

Regardless of which Ending you chose, everyone gathers in the Great Hall as Professor Black delivers a speech. Struggling to get words out, Professor Weasley steps in, and after a fond farewell and thank you to Professor Fig, you all raise a glass in his honor.

A conversation with Ominis begins, where you reveal that it was in fact Rookwood that cursed Anne, not Goblins, and he departs to get that information to both Anne and Sebastian as fast as possible. You return to your dorm as the season outside turns to spring, where you can finish up the school year by concluding any and all unfinished business, within the pages of your Field Guide, or out in the open world.

SIDE QUESTS

✧ Dissending' for Sweets

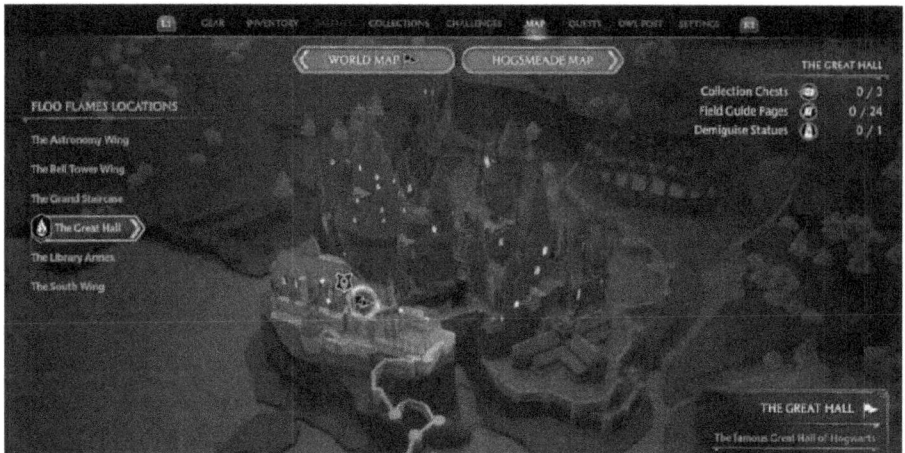

- Location: The Great Hall -> Great Hall

- Quest Giver: Garreth Weasley

- Quest Level: Level 5

- Requirement: Having Finished Main Quest: Potions Class

- Reward: Gold, Conjuration Spellcraft, 180XP

- Mission Info: It appears that Garreth Weasley wants to speak with me.

Objectives:

- Talk to Garreth Weasley

- Find the statue of the one-eyed witch

- Open the one-eyed witch statue

- Explore the secret passage

- Find a way out of the secret passage

- Find the Billywig Stings

- Return to Garreth

Starting Location: 'Dissending' for Sweets

- Talk to Garreth Weasley

  Just talk to Garreth, he wants you to get some Billywig Stings from Honeydukes.

- Find the statue of the one-eyed witch

Follow the quest marker through the door next to Garreth, go up the little staircase and you can see the statue already from far away next to a fast travel point.

■ Open the one-eyed witch statue

Now just interact with the statue to open it.

■ Explore the secret passage

Follow the given path, cast Reparo on the Elevator, then use it.

After the elevator, you will have to use Incendio on the spider webs to move on.

Now cast Levioso on the platform, then jump to move on.

Cast Levioso once more on the next platform, then jump onto the platform to move on.

Now cast Reparo, then Levioso, and jump onto the platform to move on.

Cast Incendio to destroy the spider web and move on.

Cast Reparo, then Accio on the metal latch above the platform, then Levioso on the platform, and jump onto the platform to move on.

To open the door, cast Incendio two times to light the braziers and move on.

Find a way out of the secret passage

Now enter Honeydukes by interacting with the ladder.

■ Find the Billywig Stings

Follow the path inside Honeydukes and you will see the Billywig Stings on the right side. You can cast Revelio to highlight it.

■ Return to Garreth

After you got the Stings, you can fast-travel back to Garreth or go through the hidden passage where you came from.

This finishes 'Dissending' for Sweets Side Quest in Hogwarts Legacy.

✧ Mer-Ky' Depths

- Location: The Great Hall -> Boathouse

- Quest Giver: Nerida Roberts

- Quest Level: Level 10

- Requirement: Having Finished Main Quest: Percival Rackham's Trial

- Reward: Appearances, 180XP

- Mission Info: Nerida Roberts is sitting on the docks by the Hogwarts boathouse. I should speak with her.

Objectives:

- Talk to Nerida Roberts

- Find the Leaky Caves

- Find and collect the Mermish artefact

- Return to Nerida

Starting Location: 'Mer-Ky' Depths

- Talk to Nerida Roberts

  Nerida wants you to get an artifact.

- Find the Leaky Caves

  Follow the quest marker to the Cave nearby and dive in.

- Find and collect the Mermish artefact

Inside the cave you will see a door, to open it you need to lead all three butterflies to it. For two of the three butterflies, cast Incendio to free them from the spider web, then cast Lumos and lead them to the door, cast Lumos again and they will fit into the door.

Once all three butterflies are at the door, it will open. A chest will be revealed, open the chest to get the Artifact.

- Return to Nerida

  Now just go back to Nerida.

This finishes 'Mer-Ky' Depths Side Quest in Hogwarts Legacy.

✧ A Demanding Delivery

- Location: Hogsmeade

- Quest Giver: Parry Pippin

- Quest Level: Level 5

- Requirement: Having Finished Main Quest: Flying Class

- Reward: Appearances, Gold, 180XP

- Mission Info: Parry Pippin of J. Pippin's Potions in Hogsmeade sent me an owl regarding a delivery. I should speak with him if I'm looking for some simple work.

Objectives:

- Speak With Parry Pippin

- Deliver 3 Invisible Potions to Fatimah Lawang

- Return to Parry Pippin

Starting Location: A Demanding Delivery

- Speak With Parry Pippin

Speak with Parry inside his shop in Hogsmeade, he wants you to deliver Invisible Potions to Fatimah southside of Hogwarts.

- Deliver 3 Invisible Potions to Fatimah Lawang

Follow the quest marker and travel to Keenbridge south side of Hogwarts to find Fatimah.

During the conversation, no matter what you choose here, she wants you to test the potion. So press and hold **L1 / LB** and select the potion.

Press **L1: / LB :** to use it now and talk to Fatimah again while invisible.

- Return to Parry Pippin

Now you can return to Parry and report that you delivered the potions.

This finishes A Demanding Delivery Side Quest in Hogwarts Legacy.

✧ A Friend in Deed

A Friend in Deed Walkthrough

Sirona Ryan, who works at The Three Broomsticks, wishes to meet with you. Fast travel over to Hogsmeade to learn her friend needs a favour done. Dorothy Sprottle is the friend, who lives in Upper Hogsfield. She directs you to a nearby cave where you must find Horklumps, Sirona's Letters, and a letter box. You'll find the Horklumps all along the main path, and then the letters and box are right at the end of the cave on the floor just before you exit. Return to Dorothy Sprottle to give her the Horklumps, and then give everything else to Sirona Ryan back in Hogsmeade.

A Friend in Deed Objectives

➢ Talk to Sirona Ryan in the Three Broomsticks

➢ Speak to Dorothy Sprottle in Upper Hogsfield

➢ Find and enter the cavern

➢ Return to Dorothy Sprottle in Upper Hogsfield

➢ Return the box of letters to Sirona

✧ Absonder Encounter

➢ Location: World Map -> South Hogwarts Region -> Aranshire

➢ Quest Giver: Edgar Adley

➢ Quest Level: Level 32

➢ Requirement: Having Finished Main Quest: Tomes and Tribulations

➢ Reward: Wand Handles, Gold, 180XP

➢ Mission Info: A vendor in Aranshire seems to need help with something.

Objectives:

- Talk with Edgar Adley

- Find The Absconder's Cave

- Defeat the Absconder

- Find the heirloom Milo left behind

- Return the heirloom to Edgar

Starting Location: Absonder Encounter

- Talk with Edgar Adley

    Talk to Edgar in Aranshire, he wants you to get the heirloom of his friend back from the Absconder.

- Find The Absconder's Cave

    Follow the quest marker into the Forbidden Forrest to find the cave. You have to cast Incendio to enter it.

- Defeat the Absconder

    A bit into the cave, you will find the Absconder. Cast Incendio once more to enter the arena. Dodge his attacks, then attack him, rinse and repeat. Use Incendio as much as you can. From time to time he will spawn some smaller spiders, which you should defeat or else they will be pretty annoying.

- Find the heirloom Milo left behind

    After defeating the Absconder, you will find the heirloom in a small cave behind some spider web. Use Incendio to enter it.

- Return the heirloom to Edgar

    Now just return to Edgar and give him the heirloom.

✧ Breaking Camp

- Location: World Map -> Hogsmeade Valley -> Upper Hogsfield

- Quest Giver: Claire Beaumont

- Quest Level: Level 13

- Requirement: Having Finished Main Quest: Tomes and Tribulations

- Reward: Conjuration Spellcraft, 180XP

- Mission Info: The local vendor in Upper Hogsfield, Claire Beaumont, seems worried about something. I should ask her what's wrong.

Objectives:

- Talk to Claire Beaumont

- Clear the goblin encampments (x2)

- Return to Claire Beaumont

| Starting Location: Breaking Camp

- Talk to Claire Beaumont

  Talk to Claire in Upper Hogsfield, she wants you to clear out goblin camps.

- Clear the goblin encampments (x2)

  Just follow the quest marker to the first goblin camp and defeat all enemies here.

  After defeating all enemies, follow the quest marker to the second camp and defeat all enemies there.

- Return to Claire Beaumont

  Now return to Claire and report what you did.

✧ Brother's Keeper

- Location: World Map -> Hogsmeade Valley -> Upper Hogsfield

- Quest Giver: Dorothy Sprottle

- Quest Level: Level 13

- Requirement: Having Finished Main Quest: Tomes and Tribulations

- Reward: Wand Handle, 180XP

- Mission Info: Dorothy Sprottle is concerned about a missing person. Perhaps I can help track them down

| Objectives:

- Speak with Dorothy Sprottle

- Discover what happened to Bardolph Beaumont

- Report back to Claire Beaumont

| Starting Location: Brother's Keeper

- Speak with Dorothy Sprottle

  Just talk to her, she wants you to look at what happened to Bardolph.

- Discover what happened to Bardolph Beaumont

  Follow the quest marker to a small ruin with some enemies. One of them will be Bardolph, you have to kill him.

- Report back to Claire Beaumont

  After you killed Bardolph, go to his sister by following the quest marker and report what happened.

  This finishes Brother's Keeper Side Quest in Hogwarts Legacy.

Cache in the Castle Walkthrough

You'll find this Side Quest in the Defence Against the Dark Arts Tower just outside the classroom where Charms takes place. A student called Arthur Plummly can be interacted with, and he's sending you on a little treasure hunt. He'll give you a piece of paper with some clues on it, so first, follow the objective marker to the general vicinity of the first clue. It's the large skeleton display on the first floor.

Next, head outside into the courtyard and look at the fountain to tick off the second landmark. Then cross over into the building on the other side of the courtyard and climb the flight of stairs. You're looking for a painting depicting a house in winter covered in snow. It's next to another skeleton. When you find it, cast Accio. Head inside and open the Chest for a reward. Then, return to Arthur Plummly to complete the Side Quest.

Cache in the Castle Objectives

■ Find the first landmark from Arthur's Treasure Map

■ Find the second landmark from Arthur's Treasure Map

■ Find the painting from Arthur's Treasure Map

■ Discover the painting's secret

■ Return to Arthur Plummly

✧ Carted Away

Carted Away Walkthrough

You can start this Side Quest by travelling to Lower Hogsfield and talking to Arn near the lake. He's had his possessions stolen by goblins, and you've been asked to get them back. Follow the quest marker to their hideout 500 metres away. Once you get there, kill all the goblins in the camp. Some of them can hit pretty hard and gang up on you, so be sure to have some Wiggenweld Potions in your inventory. Also, pay attention to the colour of their shields and use a corresponding Spell colour to destroy them.

Once they're all dead, go to the set of metal gates in the back and open them. The carts will return to Arn on their own, so go back to him to complete the Side Quest.

Carted Away Objectives

- Go to the Goblin Camp

- Free the carts from the encampment

- Return to Arn

✧ Crossed Wands

This Side Quest takes place in the duelling arena near the Floo Flame labelled Clock Tower Courtyard. Across a series of rounds, you'll duel different opponents. Below you'll find an explanation of all rounds that make up Crossed Wands, accessed by talking to Lucan Brattleby.

Round 1

In Round 1, you'll team up with Sebastian Sallow and fight Lawrence Davies and Astoria Crickett. They utilise yellow shields which can be beaten using Levioso. Then it's best to use Basic Cast to defeat them.

Round 2

In Round 2, you'll fight Hector Jenkins, Constance Dagworth, and Nerida Roberts. They have violet shields protecting them, but they can be broken by using Accio.

Round 3

Round 3 is unlocked after completing the Main Quest called Tomes of the Tribulations. It puts you up against Eric Northcott, Charlotte Morrison, Nellie Oggspire, and Leander Prewett. This duel introduces the red shield, which is removed using Incendio. This is the final Round and you'll be rewarded with the Crossed Wand Champions Garb for finishing it.

E-Vase-Ive Manoeuvre

E-Vase-Ive Manoeuvre Walkthrough

Madam Twiddle wants you to investigate a statue just south of Irondale, so head on over there. To activate the statue, you need to destroy all 20 vases in the nearby area. Using Revelio makes it easier to find them. Return to Madam Twiddle to complete the Side Quest.

E-Vase-Ive Manoeuvre Objectives

- Head to the statue Madam Twiddle mentioned

- Activate the statue

- Return to Althea Twiddle

✧ Flight Test

Flight Test Walkthrough

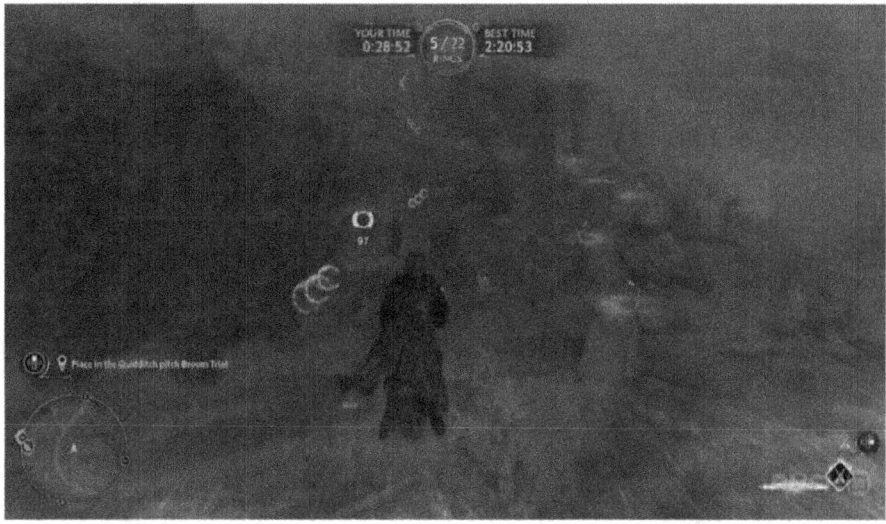

Head on over to the Quidditch Pitch to start this Side Quest, where you'll find Imelda Reyes. She has some time trials for you to beat, with the first asking you to race through 22 rings in under two minutes and 20 seconds. Beat the time and report back to Imelda Reyes. Then you'll need to go back and see Albie Weekes in Hogsmeade. If you buy his first Broom Upgrade for 1,000 Coins, you'll unlock Sweeping the Competition.

Flight Test Objectives

- Speak with Imelda Reyes at the Quidditch Pitch
- Place in the Quidditch pitch Broom Trial
- Report back to Imelda
- Return to Albie Weekes

✧ Flying Off the Shelves

Flying Off the Shelves Walkthrough

Cressida Blume needs you to head into the Library to gather her books that have begun flying. Follow the objective marker to the Library and grab the books using Accio. There are three books on the first floor and two on the second floor. Take them back to Cressida Blume to complete this Side Quest.

Flying Off the Shelves Objectives

Collect Cessida's flying books from the library

Talk to Cressida

✧ Follow the Butterflies

- Location: Hogsmeade

- Quest Giver: Clementine Willardsey

- Quest Level: Level 5

- Requirement: Having Finished Main Quest: Potions Class

- Reward: :bronze: Followed the Butterflies, Conjuration Spellcraft, 180XP

- Mission Info: I overheard Clementine Willardsey talking to herself in the Three Broomsticks. She mentioned something about a swarm of butterflies.

Objectives:

- Talk to Clementine Willardsey

- Find the butterflies in the Forbidden Forest

- Find out where the butterflies are going

- Return to Miss Willardsey

Starting Location: Follow the Butterflies

■ Talk to Clementine Willardsey

Talk to Clementine inside the Three Broomsticks, she wants you to find and follow some butterflies.

- Find the butterflies in the Forbidden Forest

Follow the quest marker to the Forbidden Forest, you will find a small swarm of Butterflies near the beginning of the Forbidden Forest.

- Find out where the butterflies are going

The Butterflies will lead you through the small river nearby and then to a chest.

Once you looted the chest you will unlock :bronze: Followed the Butterflies.

- Return to Miss Willardsey

After you found the chest, go back to Miss Willardsey.

✧ Gobs of Gobstones

Gobs of Gobstones Walkthrough

- Location: The Astronomy Wing -> Defence Against the Dark Arts Tower

- Quest Giver: Zenobia Noke

- Quest Level: Level 03

- Requirement: Having Finished Main Quest: Weasley after Class

- Reward: 180XP, 20HP, Wand Handles

- Mission Info: Zenobia Noke is fretting over her missing Gobstones.

In the Astronomy Tower near the Floo Flame labelled Professor Fig's Classroom, you'll find Zenobia Noke leaning against a wall asking for help. She needs you to find her Gobstones, of which there are six. To get them, find them and cast Accio to bring them to you. Below you'll find guides to finding all six Gobstones.

❖ Gobstone #1

In the Transfiguration Courtyard, head to the balcony area underneath the bridge. On the roof just to the left of the owl is a Gobstone.

❖ Gobstone #2

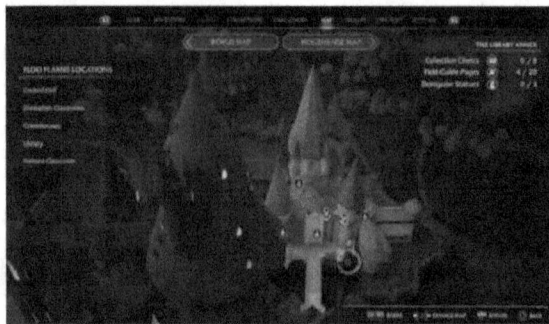 

At the top of the staircase where the Divination Classroom is, you'll find this Gobstone in the chandelier.

❖ Gobstone #3

Take the path to the right of the Floo Flame labelled Divination Classroom and look out onto the wooden beams on your left. A Gobstone is in the middle.

❖ Gobstone #4

In the Ravenclaw Tower right next to the Floo Flame, look up and you should spot this Gobstone perched near the edge of a pillar.

❖ Gobstone #5

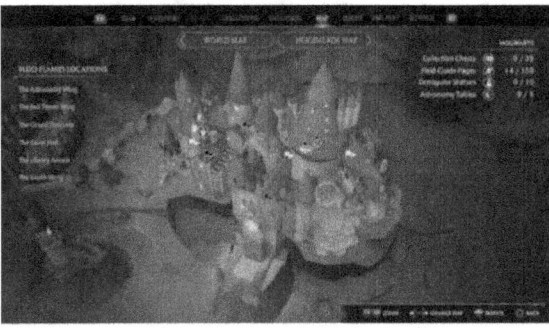 

Just outside the Trophy Room is a Gobstone near a window.

❖ Gobstone #6

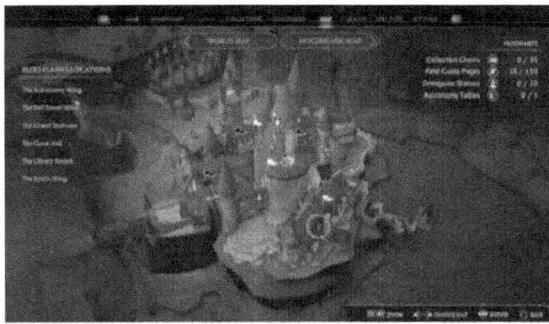 

Inside the Trophy Room, there'll be shelves of cups and trophies in the northern corner. Look up onto the top shelf to spot a gap with the Gobstone placed there.

Once you've found all the Gobstones, return them to Zenobia to complete the Side Quest.

Gobs of Gobstones Objectives

- Talk to Zenobia Noke

- Find Zenobia's Gobstones

- Return to Zenobia

✧ Interior Decorating

Interior Decorating Walkthrough

After completing the Main Quest called The Room of Requirement, you'll be able to start this Side Quest by talking to Professor Weasley. She needs you to gather Moonstone, but if you've already gone enough, you can continue immediately. The teacher wants you to decorate the room, so place five wall decorations and five floor decorations anywhere you like. Next, you'll be taught a new Spell called Altering. Use it to alter, change the colour, and adjust the size of the items in the room.

You must then customise the floor or balcony using Altering, so change it to a colour of your choosing and then speak to Deek. He'll let you change the room's ambience, and then speaking to Professor Weasley completes the Side Quest.

Interior Decorating Objectives

➤ Speak with Professor Weasley

➤ Return to Professor Weasley

➤ Conjure wall decorations

➤ Conjure floor decorations

➤ Speak to Professor Weasley in the Room of Requirement

➤ Alter the style of an item

➤ Change the colour of an item

➤ Adjust the size of an item

➤ Alter the balcony or the floor of the room

➤ Speak to Deek

✧ Kidnapped Cabbage

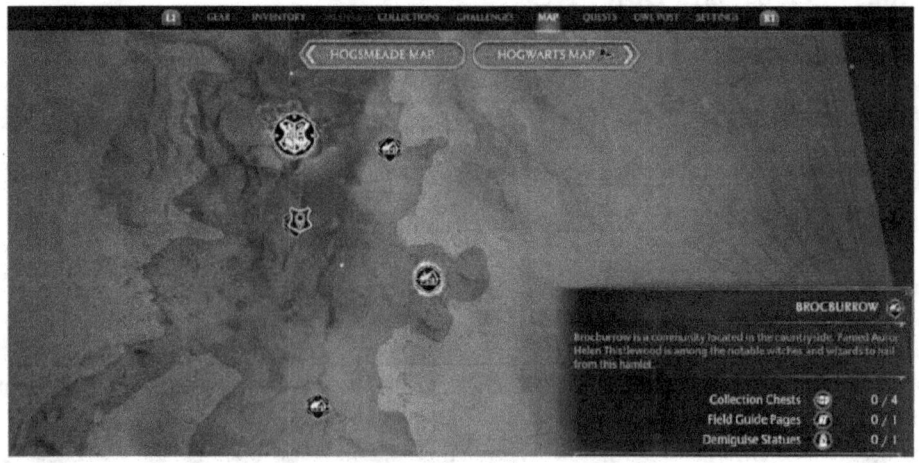

➤ Location: World Map -> Hogwarts Valley -> Brocburrow

➤ Quest Giver: Eddie Thistlewood

➤ Quest Level: Level 10

- ➢ Requirement: Having Finished Main Quest: Tomes and Tribulations

- ➢ Reward: Conjuration Spellcraft, 180XP

- ➢ Mission Info: Eddie Thistlewood in Brocburrow seems to have an urgent matter involving Chinese Chomping Cabbages that needs attending. I should speak with him.

Objectives:

- ➢ Talk to Eddie Thistlewood

- ➢ Go where the first crate was spotted

- ➢ Retrieve the cabbages

- ➢ Go where the second crate was spotted

- ➢ Retrieve the cabbages

- ➢ Deliver 4 Chinese Chomping Cabbages to Bernard Ndiaye in Feldcroft

Starting Location: Kidnapped Cabbage

- ■ Talk to Eddie Thistlewood

  Talk to Eddie in Brocburrow. He will ask you to find two crates with Chinese Chomping Cabbages.

  Go where the first crate was spotted

  Follow the Quest Marker to find the first crate.

- ■ Retrieve the cabbages

  Defeat all enemies in the camp, then collect the cabbage.

  Go where the second crate was spotted

  Follow the Quest Marker again to find the second crate, and defeat all enemies again in the camp.

- ■ Retrieve the cabbages

  After defeating all enemies, pick up the cabbage next to the tent.

  Deliver 4 Chinese Chomping Cabbages to Bernard Ndiaye in Feldcroft

  Follow the quest marker one more time to find Bernard and deliver the Chinese Chompong Cabbage to him.

✧   Like a Moth to a Frame

Like a Moth to a Frame Walkthrough

When you're in the Central Hall, look to the right of the central fountain to find a student named Lenora Everleigh investigating a painting. Talk to her and you'll be given this Side Quest. Begin by casting Lumos next to it and a picture of another place inside Central Hall will appear. To find it, go left and down the steps as if you were going outside, but instead, take a sharp left. You'll notice a moth stuck on the wall, and casting Lumos on it will allow you to take it back with you to the painting. Return it to the painting and then talk to

Lenora Everleigh again to complete the Side Quest.

Like a Moth to a Frame Objectives

- Find the location depicted in the painting

- Talk to Lenora Everleigh

✧ Solved by the Bell

- Location: World Map -> Manor Cape -> Henrietta's Hideout

- Quest Giver: Musical Map

- Quest Level: Level 13

- Requirement: Having Finished Main Quest: Tomes and Tribulations

- Reward: Appearances, 400 Galleons, 180XP

- Mission Info: The valuable item in Henrietta's Hideaway was a Musical Map. I should seek its treasure.

Objectives:

- Find the item

- Use the Musical Map to find the treasure

Starting Location: Solved by the Bell

■ Important

Before starting the quest, you need three Spells:

Incendio Finish Hecat's Assignment 1

Glacious Finish Hecat's Assignment 1

Arresto Momentum Finish Kogawa's Assignment 2

■   Find the item

To actually start the quest, you have to find the Musical Map inside Henrietta's Hideout. You can find the Hideout here.

The entrance to the hideout is below Merlins Trial, go down the stairs next to the trial and enter the hideout.

To proceed, use Incendio on the statue to your left.

This will reveal a cube, use Accio or Windardium Levioso to get the cube onto the pedestal. Then cast Incendio on the cube to the left, and Glacius spell on the cube to the right. If done right, a door behind the cubes will open.

Beat all enemies in the next room, then go up the stairs on the far left side.

On this little gangway, cast Arresto Momentum onto the wall before passing it, or else you can't pass.

You will be in a room with some enemies again, defeat them. Then go beneath this little balcony and enter the hidden room through the wall.

The Musical map is on the table in this room.

Use the Musical Map to find the treasure

Once you have the map travel to Clagmar Castle on the far southeast side of the map.

Defeat all enemies in this castle, then look for the bells inside the castle. Hit the bells with **R2/RT** in the following order to proceed.

Once you listened to the familiar melody, a chest will spawn to your left. Open it.

✧   Spell Combination Practice

Spell Combination Practice Walkthrough

The Spell Combination Practice Side Quests can be found at Crossed Wands, which is the duelling club near the Clock Tower Courtyard. To start them, speak with Lucan Brattleby there. Below you'll find an explanation of every Spell Combination Practice Side Quest in the game.

❖ Spell Combination Practice 1

Cast Accio and then Basic Cast four times

Cast Levioso and then Basic Cast four times

Cast Levioso, then Basic Cast three times, and then Accio, then Basic Cast four more times

❖ Spell Combination Practise 2

Cast Accio, then Incendio, and then Levioso

Cast Levioso, then three Basic Casts, then Accio and finish with Incendio

Cast Accio, then one Basic Cast, followed by Incendio and three Basic Casts, then Levioso followed by three more Basic Casts, then Accio, Incendio, and finish with four Basic Casts

✧ Summoner's Court: Match 1

Sweeping the Competition Walkthrough

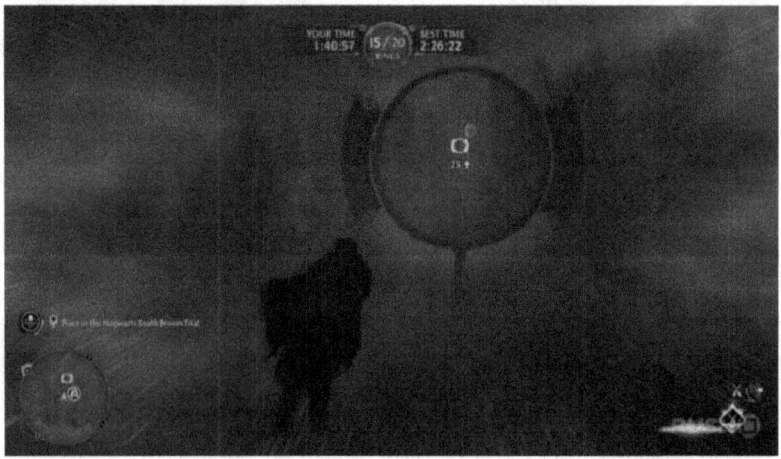

For the second Broom Trial, meet Imelda Reyes in Irondale, which is south of Hogwarts. This time, you'll need to fly through 20 rings in less than two minutes and 26 seconds. Speak to Imelda Reyes again once you've beaten her time, then head back to Albie Weekes to report the good news.

Sweeping the Competition Objectives

- Speak with Imelda Reyes

- Place in the Hogwarts South Broom Trial

- Speak with Imelda Reyes

- Return to Albie Weekes

✧ The Hippogriff Marks The Spot

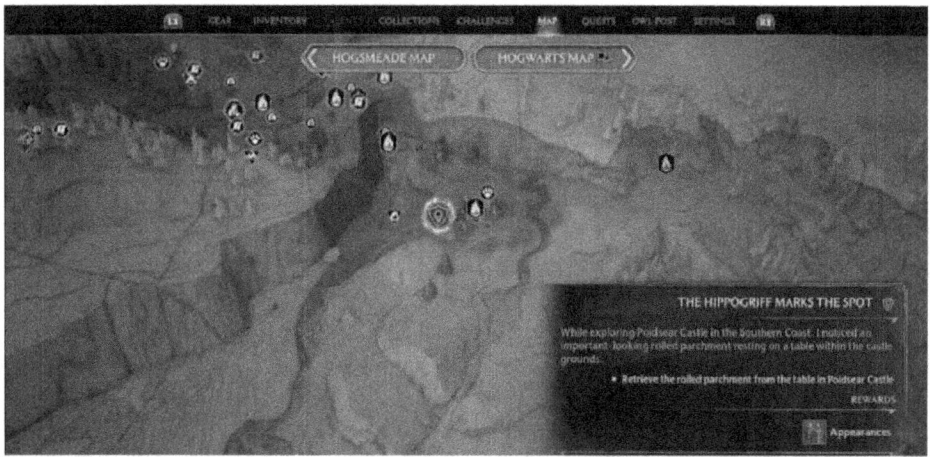

- Location: World Map -> Poidsear Coast -> Poidsear Castle

- Quest Giver: Henrietta's Map

- Quest Level: Level 13

- Requirement: Having Finished Main Quest Tomes and Tribulations

- Reward: Appearances, 180XP

- Mission Info: While exploring Poidsear Castle in the Southern Coast, I noticed an important-looking rolled parchment resting on a table within the castle grounds.

Objectives:

- Retrieve the rolled parchment from the table in Poidsear Castle

- Use Henrietta's Map to find the treasure

Starting Location: The Hippogriff Marks The Spot

- Retrieve the rolled parchment from the table in Poidsear Castle

Just pick up the rolled parchment from the table inside the castle. You will get a hint on where you have to look.

■ Use Henrietta's Map to find the treasure

Once you got the map, go to the following location.

The entrance to the hideout is below Merlins Trial, go down the stairs next to the trial and enter the hideout.

To proceed, use Incendio on the statue to your left.

This will reveal a cube, use Accio or Windardium Levioso to get the cube onto the pedestal. Then cast Incendio on the cube to the left, and Glacius spell on the cube to the right. If done right, a door behind the cubes will open.

Defeat the enemies that are in the next room, after that use Incendio on the statue in the middle of the room. Then turn the 4 highlighted flames from the picture below. This is what the hint from the map

A wall will open, revealing a Chest. Open that Chest.

✧ The Lost Astrolabe

The Lost Astrolabe Walkthrough

When you're in Lower Hogsfield, you'll spot a student standing on the dock to the east. Grace Pinch-Smedley wants you to get a family heirloom out of the Black Lake, so jump in and swim towards the objective marker. When you get to the purple circle on your mini-map, interact with the white patches of water where it's slightly choppier. You'll find some items and eventually, the item Grace is looking for. Return to shore and give Grace the item to complete this Side Quest.

The Lost Astrolabe Objectives

- Speak with the student on the dock

- Dive in the Black Lake and find the astrolabe

- Return to Grace

✧ Troll Control

- Location: World Map -> Hogwarts Valley -> Brocburrow

- Quest Giver: Alexandra Rickett

- Quest Level: Level 13

- Requirement: Having Finished Main Quest: Tomes and Tribulations

- Reward: Appearances

- Mission Info: Upon approaching Brocburrow, I overheard a woman talking about a problem with a troll. I should speak with her.

Objectives:

- Talk to Alexandra Rickett

- Go to the Troll's Den

- Dispatch Alexandra's Troll

- Return to Alexandra

| Starting Location: Troll Control

■ Talk to Alexandra Rickett

Talk to Alexandra in Brocburrow, she wants you to defeat her Troll.

Go to the Troll's Den

Follow the Quest Marker to find the Troll's Den.

■ Dispatch Alexandra's Troll

Defeat the Troll, cast Ancient Magic L1+R1 / LB+RB whenever it's ready, it deals a lot of damage. Also throw Rocks and Explosive Barrels with R1 / RB whenever you find them on the battlefield and you will defeat the Troll pretty quickly.

■ Return to Alexandra

After defeating the Troll, get back to Alexandra.

✧ Venomous Revenge

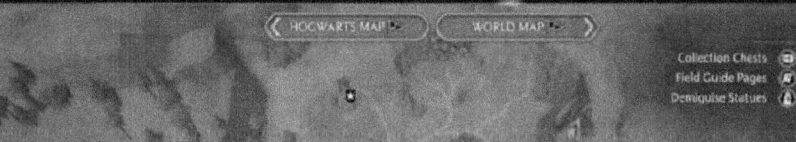

- Location: Hogsmeade

- Quest Giver: Ackley Barnes

- Quest Level: Level 15

- Requirement: Having Finished Main Quest: The Helm of Urtkot

- Reward: Gold, Appearances, 180XP

- Mission Info: A man is standing behind the Three Broomsticks, muttering angrily to himself. Should I engage?

Objectives:

- Talk to the muttering man behind the Three Broomsticks

- Find Alfred Lalley's Cellar

- Find Lawley's Venomous Tentacula

- Return to Ackley Barnes

Starting Location: Venomous Revenge

Talk to the muttering man behind the Three Broomsticks

Talk to Alfred behind the Three Broomsticks, he wants you to get Venomous Tentacula.

■ Find Alfred Lalley's Cellar

Follow the quest marker to find the cellar, it's near where you started this quest.

■ Find Lawley's Venomous Tentacula

Inside the cellar, cast Disillusionment to be invisible. If Alfred sees you you have to restart this section. Once you are inside the cellar and invisible, stay to the right side and go through the door on the right side.

After you passed the door you will be in a room with some enemies, you can be visible again. After you defeated the enemies go up the small set of stairs to find the Venomous Tentacula.

■ Return to Ackley Barnes

After you picked up the Tentacula, cast Disillusionment to get invisible and follow the Quest Marker again to get out of the Cellar. Again, don't let Alfred see you. Return to Ackley now.

✧ Venomous Valour

Venomous Valour Walkthrough

Venemous Valour is started by talking to Duncan Hobhouse on the bottom floor of the Defence Against the Dark Arts Tower. He wants you to find a Hidden Herbology Corridor, so follow the objective marker there. It'll lead you to an area outside where you can clearly spot a doorway blocked by vines. To clear them, cast Incendio. Once inside, cast Lumos to cross the Devil's Snare. Continue to cast Lumos as you work your way down the corridor, eventually finding a Giant Venomous Tentacula. Approach it and pick a leaf off, then leave via the ladder to your right. Go back to Duncan Hobhouse with the leaf in hand and you'll complete this Side Quest.

Venemous Valour Objectives

- Find and enter the Hidden Herbology Corridor

- Obtain proof for Duncan

- Collect a Giant Venomous Tentacula leaf

- Return to Duncan

ASSIGNMENTS

✧ Madam Kogawa's Assignment 1

Madam Kogawa's Assignment 1 Walkthrough

To complete this Assignment, you simply need to pop five balloons over Hogsmeade Station and then another five around the Quidditch Pitch. If you're having trouble finding them, refer to our all Balloons Locations guide. Return to Madam Kogawa after popping them all to complete the Assignment and learn Glacius.

Madam Kogawa's Assignment 1 Objectives

- Pop balloons over Hogsmeade Station

- Pop balloons around the Quidditch Pitch

- Return to Madam Kogawa

✧ Madam Kagowa's Assignment 2

Madam Kagowa's Assignment 2 Walkthrough

It's back to balloon popping for this Assignment, this time around the Spires and Keenbridge Tower. Simply fly to both places on a Broomstick and pop five balloons at each location. Go back and see Madam Kagowa at the castle to unlock Arresto Momentum.

Madam Kagowa's Assignment 2 Objectives

Practise flying near the Spires

Practise flying near Keenbridge Tower

Return to Madam Kagowa

✧ Professor Garlick's Assignment 1

To complete this Assignment, you're required to use a Venomous Tentacula and then use a Mandrake on multiple enemies. To get these plants, head to Dogweed and Deathcap in Hogsmeade and buy them. Don't bother buying them as seeds, the Combat Tools will do. Both can be equipped using the L1 radial menu, and you literally just need to use a Venomous Tentacula to complete that objective.

As for the Mandrake, find a group of enemies and use the plant and it'll scream at them. To complete the Assignment, attend a Herbology class during the day and you'll be taught Wingardium Leviosa.

✧ Professor Hecat's Assignment 1

Professor Hecat's Assignment 1 Walkthrough

After completing the Main Quest called The Locket's Secret, you'll be immediately handed this Assignment. Follow the objective marker to find Professor Hecat in her classroom. She'll set you two tasks: win two rounds of Crossed Wands and complete a round of spell combination with Lucan Brattleby. Again, follow the objective marker to your first objective, which is the duelling club. Win two duels, and if you haven't already done them, you can combine this task with the Side Quest called Crossed Wands.

After that, speak to Lucan Brattleby again to do a round of Spell Combination Practice. Again, this is a Side Quest if you haven't already completed it. Once the two tasks are complete, return to Professor Hecat to complete the Assignment and unlock Incendio.

Professor Hecat's Assignment 1 Objectives

Report to Professor Hecat

Win two rounds of Crossed Wands

Complete a round of spell combination practice with Lucan Brattleby

Return to Professor Hecat

✧ Professor Garlick's Assignment 2

Location: The Library Annex -> Greenhouses

Quest Giver: Professor Garlick

Quest Level: Level 16

Requirement: Having Finished Professor Garlick's Assignment 1

Reward: Flipendo, 150XP

Mission Info: Professor Garlick wants me to field test Chinese Chomping Cabbages, a Mandrake, and a Venomous Tentacula all at once. She also wants me to grow and harvest a Fluxweed plant. Should I need any seeds or plants, I could visit The Magic Neep or Dogweed and Deathcap. My Field Guide will not guide me through this assignment, but my map could still prove useful.

■ Objectives:

Grow and harvest Fluxweed

Acquire all three combat plants and use them simultaneously

Return to Professor Garlick

■ Starting Location: Professor Garlick's Assignment 2

Grow and harvest Fluxweed

For this, go to Hogsmeade and enter Tomes and Scrolls on the South Side of Hogsmeade. Buy Potting Table with a large pot spellcraft for 1000 Galleons. Now got to the Magic Neep on the west side of Hogsmeade. Buy the Fluxweed Seed. Now go to the Room of Requirement, place the Potting Table somewhere in the room, and grow the Fluxweed, it takes 15 real-time minutes to grow the Fluxweed.

Acquire all three combat plants and use them simultaneously

For this, you need to use Venomous Tentacula, MandrakeGo, and Chinese Chomping Cabbage at the same time. You don't need to be in combat to use them. Gto Dogweed and Deathcap on the North Side of Hogsmeade. You can buy all three plants directly for a total of 1400 Galleons. Or you can buy the Seeds for each plant, which costs 2450 Galleons in total. It's more expensive but you need those anyway for :silver: Put Down Roots. Once you have all three plants, just press and hold :l1: / :lb: and select a plant then release it to equip it. Press and hold :l1: / :lb: again to cast it. Repeat this for the other two plants.

Return to Professor Garlick

Go to The Library Annex -> Greenhouses and talk to Professor Garlick. She'll teach you Flipendo now.

✧ Professor Hecat's Assignment 2

Professor Hecat's Assignment 2 Walkthrough

Unlike the two previous Assignments, this particular one doesn't have you report to a teacher to start it. Instead, you're working towards it during combat situations. You need to avoid enemy attacks by dodge rolling out of the way of them 10 times, and cast Incendio on enemies five times. A great place to do both these things is in the Side Quest called Crossed Wands: Round 3. You unlocked it at the same time as this Assignment. Repeat that Side Quest a few times — purposefully failing if you haven't completed both objectives by the end — and you'll have this Assignment finished in no time.

Once both objectives have been knocked off, head to a Defence Against the Dark Arts Class during the daytime and talk to Professor Hecat after. She'll teach you Expelliarmus.

Professor Hecat's Assignment 2 Walkthrough

Successfully avoid enemy attacks by dodge rolling

Cast Incendio on enemies

Attend Defence Against the Dark Arts Class during the day

Return to Professor Hecat

✧ Professor Howins's Assignment

Professor Howin's Assignment Walkthrough

In order to unlock a Spell called Bombarda, we need to catch a Diricawl and a Giant Purple Toad with the nab-sack. Both objectives will be marked on the map so you shouldn't have any trouble finding them. Once you've got them in your nab-sack, go to Beasts Class during the day. Talking to Professor Howin after class unlocks Bombarda.

Professor Howin's Assignment Objectives

Acquire a Diricawl with the nab-sack

Acquire a Giant Purple Toad with the nab-sack

Attend Beasts Class during the day

Return to Professor Howin

✧ Professor Ronen's Assignment

Professor Ronen's Assignment Walkthrough

Professor Ronen's Assignment is picked up after completing the Main Quest called Weasley After Class. You'll need to complete it in order to progress and unlock Welcome to Hogsmeade. Leave the Transfiguration Classroom and follow the objective marker out into the courtyard, where Professor Ronen will be standing next to the statue.

In order to learn a new Spell, the professor gives you two tasks to complete. You need to collect flying pages in two different places. To do so, target them and then cast Accio. Once you've got them both, head back to Professor Ronen in the courtyard.

You'll now be given the tutorial for Reparo. Simply complete it and then repair the statue nearby to complete your first Assignment. Doing so will unlock the next bit of the story: Welcome to Hogsmeade.

Professor Ronen's Assignment Objectives

Report to Professor Ronen

Collect the flying page near the broken statue

Collect the flying page in the Defence Against the Dark Arts Tower

Return to Professor Ronen

✧ Professor Onai's Assignement

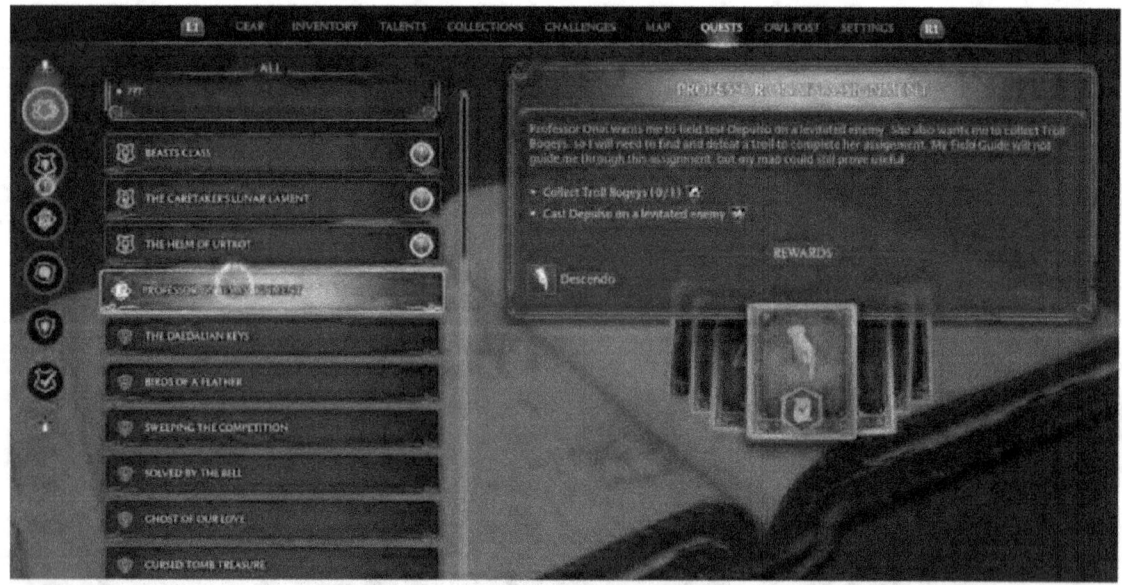

- Location: The Library Annex -> Divination Classroom

- Quest Giver: Professor Onai

- Quest Level: Level 15

- Requirement: Having Finished Main Quest: Percival Rackham's Trial

- Reward: Descendo, 150XP

- Mission Info: Professor Onai wants me to field test Depulso on a levitated enemy. She also wants me to collect Troll Bogeys, so I will need to find and defeat a troll to complete here assignment. My Field

Guide will not guide me through this assignment, but my map could still prove useful.

- Objectives:

Collect Troll Bogeys x1

Cast Depulso on a levitated enemy

Attend Divination Class during the day

Return to Professor Onai

- Starting Location: Professor Onai's Assignment

Collect Troll Bogeys x1

You just have to defeat a troll, you can find one on the south side of the map, look at the picture below.

Defeat the troll and collect the Bogey from the ground.

Cast Depulso on a levitated enemy

For this cast Levioso on a small enemy so that it floats in the air, then cast Depulso on him. If you defeated the troll from above, just go to the nearby castle, there are some enemies where you can do this.

Attend Divination Class during the day

Go back to the castle The Library Annex -> Divination Classroom and attend the Class.

Return to Professor Onai

After the class talk to Professor Onai, she will teach you Descendo.

✧ Professor Sharp's Assignment 1

Requirement: having completed Main Quest: Jackdaw's Rest

Reward: 150 XP, Depulso

Quest Info: As directed by Professor Sharp, I successfully field tested the Maxima, Edurus, and Focus Potion. In return, Professor Sharp instructed me on how to cast Depulso, the Banishing Charm that pushes opponents way with tremendous force.

- Objectives:

  - Acquire and Use a Focus Potion

  - Acquire Maxima and Edurus Potion, and use them simultaneously

  - Attend Potions Class During the Day

  - Return to Professor Sharp

Starting Location: Professor Sharp's Assignment 1

We've collected the three missing pages and made our way back to Hogwarts through the Dungeons. Professor Sharp now wants to teach us Depulso, but to do so, we must first complete an assignment for him.

- Acquire and Use a Focus Potion

Warp to West Hogsmeade floo flame and enter J. Pippin's Potions. Talk to him and buy a Focus, Edurus and Maxima Potion. These cost 500, 300, and 300 Galleons respectively. While you're at it, sell any unused gear to make room for more. Now press and hold :l1: and use the Focus Potion.

Acquire Maxima and Edurus Potion, and Use them Simultaneously

Repeat the same for the Maxima and Edurus Potions.

- Attend Potions Class During the Day

Now warp to Potions Classroom in the Library Annex at Hogwarts Castle and proceed to the next quest marker. If you need to advance time, press :touchpad: to open the map and :r3: to advance to daytime. Watch the cutscene.

- Return to Professor Sharp

Speak to Professor Sharp to start learning Depulso, the Banishing Charm.

As usual, steady your wand with the left stuck and guide it along the symbol's path to learn the spell. Don't let go of the left stick and remember to press the corresponding button to gain speed and lose the red spark. If you want to practice Depulso, there are some flying book in the classroom.

✧ Professor Sharp's Assignment 2

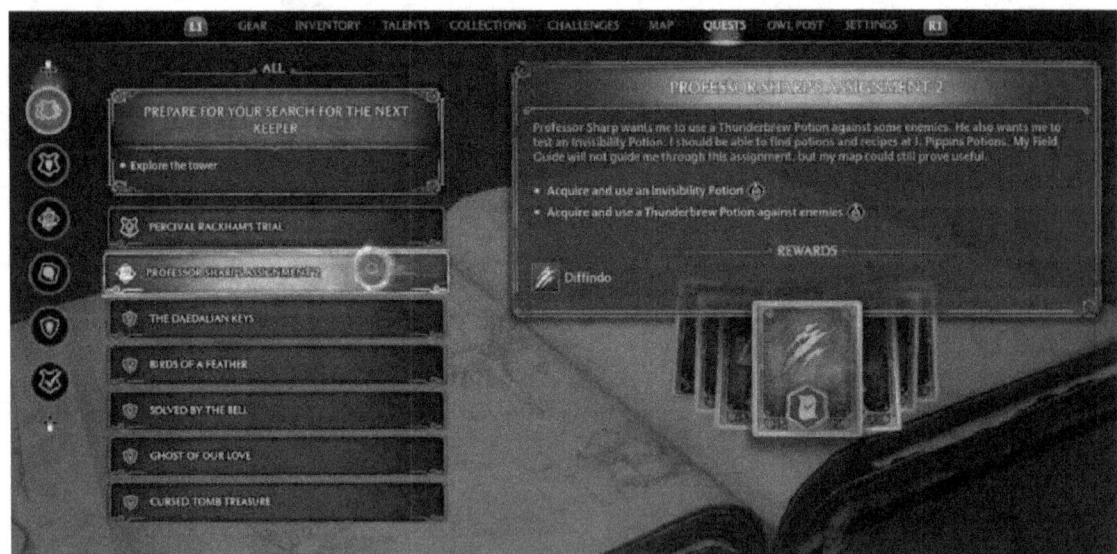

- Location: The Library Annex -> Potions Classroom

- Quest Giver: Professor Sharp

- Requirement: Having Finished Professor Sharp's Assignment 1

- Reward: Diffindo, 150XP

Mission Info: Professor Sharp wants me to use a Thunderbrew Potion against some enemies. He also wants me to test an Invisibility Potion. I should be able to find potions and recipes at J. Pippins Potions. My Field Guide will not guide me through this assignment, but y map could still prove useful.

- Objectives:

  - Acquire and use an Invisibility Potion

  - Acquire and use a Thunderbrew Potion against enemies

  - Return to Professor Sharp

Starting Location: Professor Sharp's Assignment 2

Acquire and use an Invisibility Potion

You can buy the Potion from J. Pippin's Potions in Hogsmeade. You can buy the potion directly or you can buy the recipe and brew it yourself. It's a bit more expensive, but it's needed anyway for:silver: Put Down Roots.

Acquire and use a Thunderbrew Potion against enemies

You can buy the Potion from J. Pippin's Potions in Hogsmeade. You can buy the potion directly or you can buy the recipe and brew it yourself. It's a bit more expensive, but it's needed anyway for :silver: Put Down Roots.

- Return to Professor Sharp

Now go to The Library Annex -> Potions Classroom and talk to Professor Sharp, he will teach you Diffindo now.

✧ Professor Sharp's Assignment 1

Professor Sharp's Assignment 1 Walkthrough

This Assignment builds upon your first Potions Class by making you brew more potions. It's possible you don't have the Focus Potion, Maxima Potion, or Edurus Potion recipes in your inventory yet. If you don't, visit J. Pippin's Potions in Hogsmeade and buy them. You can also buy all of the ingredients listed below from the same shop. Now return to the Potions Classroom in The Library Annex and use your Potions Station to make the potions you need.

For the Focus Potion, you'll need Lacewing Flies, Fluxweed Stem, and a Dugbog Tongue. Brew it, wait the 60 seconds, then drink it using the L1 button.

For the Edurus Potion, you'll need Ashwinder Eggs and Mongrel Fur. For the Maxima Potion, you'll need Leech Juice and a Spider Fang. You must drink either one of the potions, and then immediately select the other one and drink that too while the first potion is still active to complete the objective.

With both tasks complete, attend a Potions Class during the day and you'll finish the Assignment and unlock a new Spell called Depulso.

Professor Sharp's Assignment 1 Objectives

Acquire and use a Focus Potion

Acquire Maxima and Edurus Potions, and use them simultaneously

✧ Professor Weasley's Assignment

- Location: The Astronomy Wing -> Transfiguration Courtyard

- Quest Giver: Professor Ronen

- Quest Level: Level 3

- Requirement: Having Finished Main Quest: Defence Against Charms Class

- Reward: Reparo Spell, 150XP

- Mission Info: Professor Ronen would like to speak with me about some additional assignments.

Objectives:

- Report to Professor Ronen

- Collect the flying page near the broken statue.

- Collect the flying page in the Defence Against the Dark Art Tower

- Return to Professor Ronen

Starting Location: Professor Ronen's Assignment

- Report to Professor Ronen

  Talk to Professor Ronen in the courtyard.

- Collect the flying page near the broken statue.

  The flying page flies around the opposite of Professor Ronen. Use Accio to get it.

- Collect the flying page in the Defence Against the Dark Art Tower

  After getting the flying page near the broken statue, follow the quest marker and go into the Defence Against the Dark Art Tower next to the courtyard. This page flies around the staircase area. Use Accio to get it.

- Return to Professor Ronen

  Now that you have both pages, go to Professor Ronen in the courtyard. You will now learn Reparo.

DEMIGUISE STATUES LOCATIONS

In Hogwarts Legacy, there are many types of locks that will block your path as you seek to explore the depths of Hogwarts, as well as access to certain buildings in chests also found in Hogsmeade Village and the Highlands. As you learn the Alohomora unlocking spell from the groundskeeper, Mr. Moon, you'll find out about Demiguise Statues - small statues of creatures holding glowing moons that can be obtained only when they shimmer at night. By finding these Demiguise Statues, Mr. Moon will teach your more powerful versions of Alohomora to unlock level 2 and level 3 locks. This page includes locations for all the Demiguise Statues you can find in every region.

Note that even though you can spy a few Demiguise statues early out in the open, none of them can be interacted with until you learn the Alohomora spell. This will happen after you complete The First Trial quest and the season turns to fall, at which point Mr. Moon will ask for your assistance and teach you the unlocking spell.

Though the quest will have you grab 3 Demiguise statues around Hogwarts Castle, none of them count towards the tracked total to be found, and you'll only get one Demiguise Moon to trade to the groundskeeper towards learning advanced versions of the spell.

- You will need 9 Demiguise Moons to learn Rank 2 of Alohomora

- You will need 9 more Demiguise Moons to learn Rank 3 of Alohomora (it seems some users report the quest objective requiring 13).

✧ Demiguise Statue Location Sections

Below you will find all known locations of Demiguise Statues. Some may be found out in the open, while others are hiding behind locked doors that require a certain level of Alohomora to breach. The statue locations below are divided up into Hogwarts, Hogsmeade, and the Highlands, in the order you can find them (ones that do not require lockpicking first).

Remember, it must be night time before the Demiguise Statue can be picked up, and only after completing Mr. Moon's first quest. Despite his warnings, you won't have to worry about curfew or patrols at night after his quest. See our Hogwarts Legacy Map for more collectible locations.

If you need to change the time of day, open up the map and press the indicated button (F or R3) to wait from day to night).

✧ Hogwarts Demiguise Statue Locations

There are 10 Demiguise Statues in total to find in Hogwarts Castle, some of which will require a certain level of Alohomora to locate.

Hogwarts Demiguise Statue Location - Professor Fig's Office

This Demiguise Statue may be seen early on in the game, as it is located prominently in Professor's Fig's Office in the Defence Against Dark Arts Tower in the Astronomy Wing. Use the Professor Fig's Floo Flame to enter his office behind the classroom at night, and grab the Demiguise Statue off a desk at the back of the room.

Hogwarts Demiguise Statue Location - Divination Classroom

This Demiguise Statue may also be seen early on in the game, as it is located in the Divination Classroom - though it's easy to miss the entrance.

You can find it as you enter the Library Annex through the Viaduct Entrance, by looking to the Northeast for a large spiral staircase. Take the stairs all the way, and you'll finally reach a trap door where a rung ladder will descend automatically.

Take the ladder up, and look near the back of the room for the Demiguise Statue hiding away.

Hogwarts Demiguise Statue Location - Restricted Section Library

It's possible to spot this Demiguise Statue long before you can grab it, as you'll likely pass by it during the main quest to enter the Restricted Section of the Library in the Library Annex wing.

Once you have the ability to claim it, return to the Library and enter the Restricted Section (Which should no longer be guarded - even at night).

Head down to the bottom floor of the Restricted Section, and cross through the various rooms until you reach the hallway leading to the storage area past an Eyeball Statue, and the Demiguise Statue will be on a small end table near the door.

Hogwarts Demiguise Statue Location - Great Hall Room

This Demiguise Statue is one of the early statues to be hiding behind a locked door in Hogwarts - but completing Mr. Moon's quest will allow you access to it.

Travel to the Great Hall Dining room, and look for a level 1 locked door in the southwest corner near the professor's seating area.

Unlock the door, and you'll find the Demiguise Statue in the room with a Collection Chest you can also grab.

Hogwarts Demiguise Statue Location - South Wing Bathroom

This Demiguise Statue is one of the early statues to be hiding behind a locked door in Hogwarts - but completing Mr. Moon's quest will allow you access to it.

Travel to the Faculty Tower of the South Wing, and head down the stairs to the ground floor of Gryffindor Tower. Move through the hall until you reach the locked bathroom where Peeves is said to cause chaos, and unlock it with Alohomora.

Inside, open each of the bathroom stalls to find one is actually hiding a crawlspace into the boiler room. It is here that you can find a Demiguise Statue hidden away.

Hogwarts Demiguise Statue Location - North Hall Dungeons

This Demiguise Statue is one of the early statues to be hiding behind a locked door in Hogwarts - but completing Mr. Moon's quest will allow you access to it.

Travel to the Bell Tower Courtyard Floo Flame, and head into the North Hall before going down the stairs into the dungeons. Climb down and go past the sleeping dragon statue to find two locked doors - one of which has a level 1 lock and a makeshift drawbridge.

Enter the Muggle Studies Classroom, and immediately turn right to look to the side of the door to find the hiding Demiguise Statue in the corner, and add it to your collection.

Hogwarts Demiguise Statue Location - Bell Tower Ramparts

This Demiguise Statue is one of the early statues to be hiding behind a locked door along the Hogwarts Grounds - but completing Mr. Moon's quest will allow you access to it.

Head outside the Bell Tower Courtyard Floo Flame and go West towards the ramparts gate, and look along the right ramparts wall for a locked door under a staircase up the ramparts. Open it with Alohomora, and inside you'll find another Demiguise Statue.

Hogwarts Demiguise Statue Location - Long Gallery

This Demiguise Statue is one that will require you to have Level 2 of the Alohomora Spell, which means you'll need to find other Demiguise Moons to give to Mr. Moon first in order to learn the next level of the spell.

From the Bell Tower Courtyard Floo Flame, turn around unlock the level 1 door into the Long Gallery that connects to the Library Annex. As you move halfway through the long hall, you'll find a level 2 door on the right. Unlock it, and inside you'll find the Demiguise Statue.

✧ Hogsmeade Demiguise Statue Locations

There are 9 Demiguise Statues in total to find in the wizarding village of Hogsmeade, some of which will require a certain level of Alohomora to locate.

Hogsmeade Demiguise Statue Location - Tomes and Scrolls

You may spot this Demiguise Statue long before you can actually grab it, as it's located in a shopkeeper's home you'll meet on your first visit to Hogsmeade. As you enter the village from the southern bridge, look on the left for the Tomes and Scrolls shop, and wait until night to enter.

Luckily, the proprietor doesn't really care that you're sneaking about after hours, so walk behind Thomas Brown's counter and into his bedroom, where you'll find the Demiguise Statue glowing on a dresser at night.

Hogsmeade Demiguise Statue Location - Hog's Head Inn

This is another Demiguise Statue you may be able to spot long before you can actually grab it, as it's located out in the open - in the Hog's Head Inn, which you can find on the far northwestern edge of the village - west of the West Hogsmeade Floo Flame.

As you enter, walk behind the counter to the back room on the right, and in the tiny space leading to a door to the docks, you won't be able to miss the glowing Demiguise Statue on a stack of crates.

Hogsmeade Demiguise Statue Location - Dervish and Bangs

This is another Demiguise Statue you may be able to spot long before you can actually grab it, as it's located out in the open - at the building that houses both Gladrags Wizardwear, and Dervish and Bangs, which you can find by central Hogsmeade Square, west of the North Hogsmeade Floo Flame.

Walking inside either door, you'll find the Demiguise Statue on the main counter in the Dervish and Bangs side of the shop, next to an enchanted smithing hammer banging away.

Hogsmeade Demiguise Statue Location - Three Broomsticks

This Demiguise Statue is one of the early statues to be hiding behind a locked door in Hogsmeade - but completing Mr. Moon's quest will allow you access to it. It's located at the Three Broomsticks operated by Sirona Ryan.

Travel to the Three Broomsticks pub in the center of Hogsmeade down its main street, and climb all the narrow staircases inside to the top floor, where you'll find a level 1 locked door that you can use Alohomora on to unlock.

Once inside, you'll find the Demiguise Statue inside the small bedroom here on a central table, right next to a Field Guide Page and Collection Chest.

Hogsmeade Demiguise Statue Location - Spire Street House

This Demiguise Statue is one of the early statues to be hiding behind a locked door in Hogsmeade - but completing Mr. Moon's quest will allow you access to it. You can find it inside a locked house on Spire Street.

It's located in a small but tall building just East of the Three Broomstick, taking the southern road that runs past the establishment. You can use Alohomora on the lock and help yourself to the Demiguise Statue on the cramped floor above.

Hogsmeade Demiguise Statue Location - River's Edge House

This Demiguise Statue is one of the early statues to be hiding behind a locked door in Hogsmeade - but completing Mr. Moon's quest will allow you access to it. You can find it inside a locked house on the edge of the river near the northern part of Hogsmeade.

Look for a solitary building across from the Brood and Peck animal shop above the Central Hogsmeade Square, and use Alohomora on the level 1 lock on the door by the water. Inside, you'll find the Demiguise Statue on a mantle above the fireplace near the door.

Hogsmeade Demiguise Statue Location - Easternmost House

This Demiguise Statue is one of the early statues to be hiding behind a locked door in Hogsmeade - but completing Mr. Moon's quest will allow you access to it. You can find it inside a locked house on the far east edge of Hogsmeade.

Run around behind the large Honeydukes Sweet Shop to find two small houses behind it, and approach the north-most of the two to unlock its level 1 door. Inside you can find the Demiguise Statue on display on a table under a portrait of a wizard.

Hogsmeade Demiguise Statue Location - Lower High Street

This Demiguise Statue is one of two statues that will require you to gain Alohomora Rank 2 to be able to find it, meaning you'll have to find 9 Demiguise Statues elsewhere first. The statue is hiding on the corner of Hog's Head Street, between the Tomes and Scrolls and Ollivander's Wand Shop.

Look for a home on a small street leading towards the Hog's Head Inn that has a level 2 lock on it, and use Alohomora rank 2 to get inside the house, where you'll find the Demiguise Statue on the second floor tucked away.

Hogsmeade Demiguise Statue Location - Western River's Edge Home

This Demiguise Statue is one of two statues that will require you to gain Alohomora Rank 2 to be able to find it, meaning you'll have to find 9 Demiguise Statues elsewhere first. The statue is hiding in a building along the River's Edge on the western side of Hogsmeade.

Look for the house just to the right of Pippin's Potions Shop, North of the West Hogsmeade Floo Flame, and use rank 2 of the Alohomora unlock spell to gain entrance to the house. You'll find the Demiguise Statue on the top floor by the bed.

✧ The Highlands Demiguise Statue Locations

There are 11 Demiguise Statues in total to find in open world regions of The Highlands, some of which will require a certain level of Alohomora to locate.

South Hogwarts Region Demiguise Statue Location - South Hogsfield

This Demiguise Statue can be obtained as soon as you have undertaken the quest from Mr. Moon to find more of the Demiguise Moons, and doesn't technically require unlocking any doors to find - so you may have stumbled upon it early without realizing it.

It's located at the hamlet of South Hogsfield, south along the road from Hogwarts and where the main questline takes you to find your first Merlin Trial.

At the hamlet, look for the larger unlocked home on the North side, and step inside. You'll find the Demiguise Statue when turning around, as it will be placed on the floor by the front door.

South Hogwarts Region Demiguise Statue Location - Aranshire

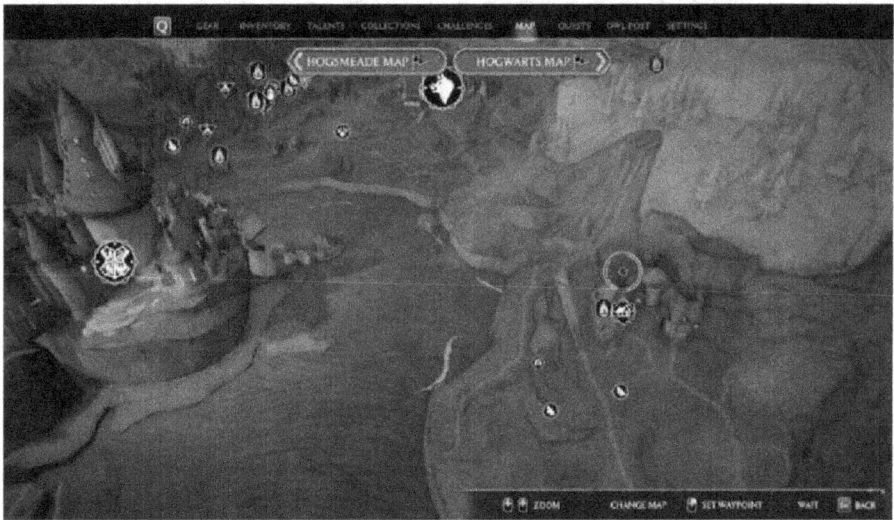

You can locate this Demiguise Statue at the hamlet of Aranshire, located East across the giant lake from Hogwarts.

It sits inside a home with a level 1 lock, so you'll need to complete Mr. Moon's initial quest before you can find its location.

Enter the hamlet and go inside the farthest house to the North in the small village area, and unlock the door. Once inside, climb to the second floor, where you'll find the Demiguise Statue tucked away.

Hogsmeade Valley Demiguise Statue Location - Upper Hogsfield

This Demiguise Statue can be found in the hamlet of Upper Hogsfield, located due North of Hogsmeade

Village, but you'll need to gain the Alohomore Unlocking Spell before you can reach it.

Once you have it, look along the central road in the small hamlet for a building in the middle, on the road towards the large building with the barrels inside, and unlock the door with a spell to find a Collection Chest inside behind a thin wall, with a Demiguise Statue in front.

North Ford Bog Demiguise Statue Location - Pitt-Upon-Ford

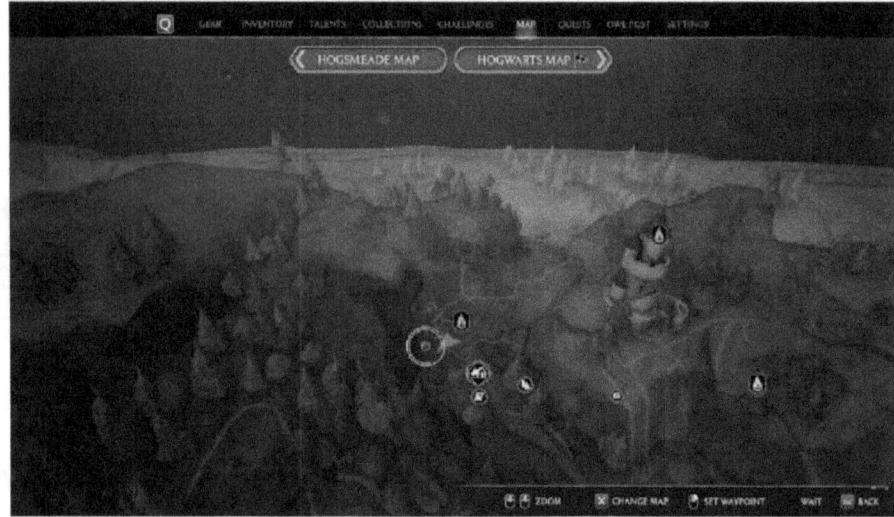

For this Demiguise Location, you won't need the unlocking spell, but you will need to know where to look. You'll find it far in the north in the North Ford Bog, at the northernmost hamlet of Pitt-Upon-Ford, which is located just to the west of the large San Barkar's Tower easily spotted from the air.

Once in the hamlet, make your way northwest across the left side of the river and you'll find a solitary tall house south of the Floo Flame, with a red carriage out front. The doors are unlocked, so you can enter any time you want.

Climb to the very top of the tall house, and you can locate the Demiguise Statue in the corner of the loft next to the bed.

Hogwarts Valley Demiguise Statue Location - Brocburrow

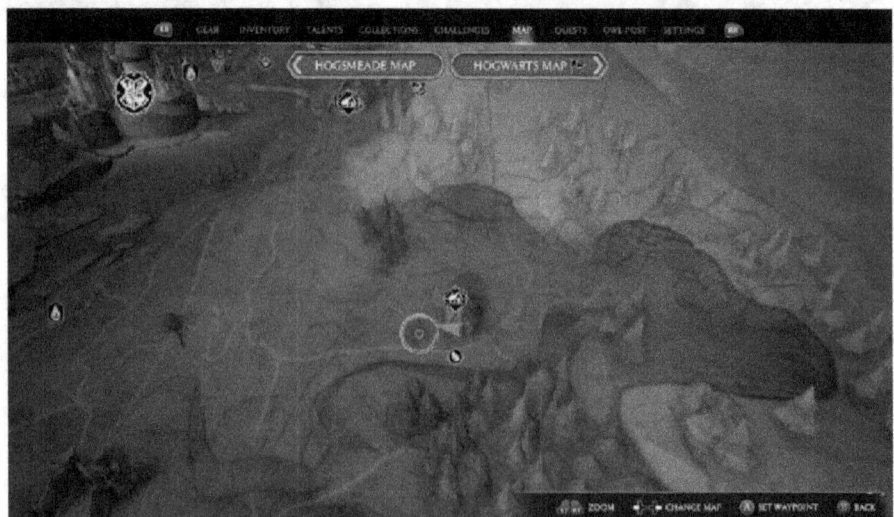

For this Demiguise Location, you won't need the unlocking spell, but you will need to know where to look. You'll want to travel to the far eastern side of Hogwarts Valley to a small hamlet nestled up in the hills called Brocburrow.

Once you arrive in the hamlet, find the Floo Flame and start walking West until you hit the Merlin Trial platform in the middle of the village. Just northwest across from the Merlin Trial is a small unlocked house where your prize awaits.

Enter the house with no trouble, and turn around after going through the door - you can find the Demiguise Statue here along a floor shelf that's easily concealed.

Hogwarts Valley Demiguise Statue Location - Keenbridge

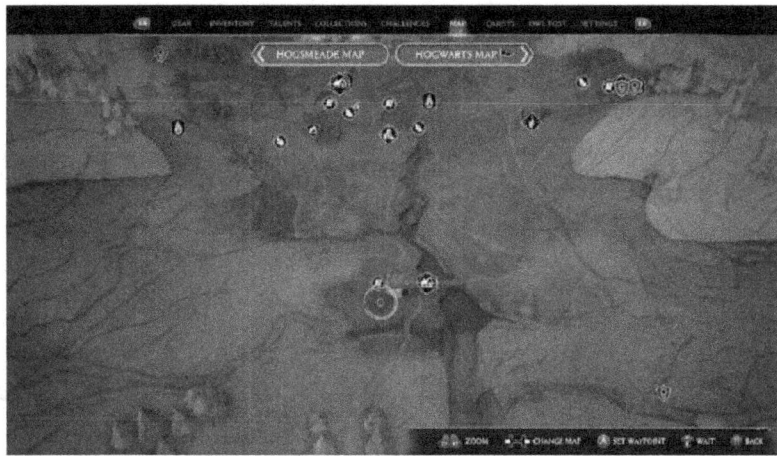

You can locate this Demiguise Statue at the hamlet of Keenbridge in the center of Hogwarts Valley, along its main river leading from the great lake. You will need to at least have the level 1 unlocking spell Alohomora to reach this Demiguise as it is behind a locked door.

From the Keenbridge Floo Flame, head west through the hamlet square past the shopkeeper Fatimah, and look for a tall house with a pumpkin patch and a large clothesline angling down from its lower roof. Use the lockpicking spell on the door to gain access.

Once inside, the Demiguise Moon will be right in front of you on a table by the stairs and a giant vase.

Feldcroft Region Demiguise Statue Location - Irondale

For this Demiguise Location, you won't need the unlocking spell, but you will need to know where to look. You'll find the small hamlet in the southwest corner of the Feldcroft Region bordering the Hogwarts Valley, located along a giant windmill that's hard to miss.

The Demiguise Statue hides in a house that has no locks, and you can locate it between the Floo Flame chimney and the giant windmill, as there is only one house between them, and a questgiver, Althea Twiddle, is sitting outside.

Enter the open home, and look carefully - as the Demiguise is hiding behind a counter along the floor, and is easy to miss unless you inspect the home's floor carefully.

Feldcroft Region Demiguise Statue Location - Feldcroft Village

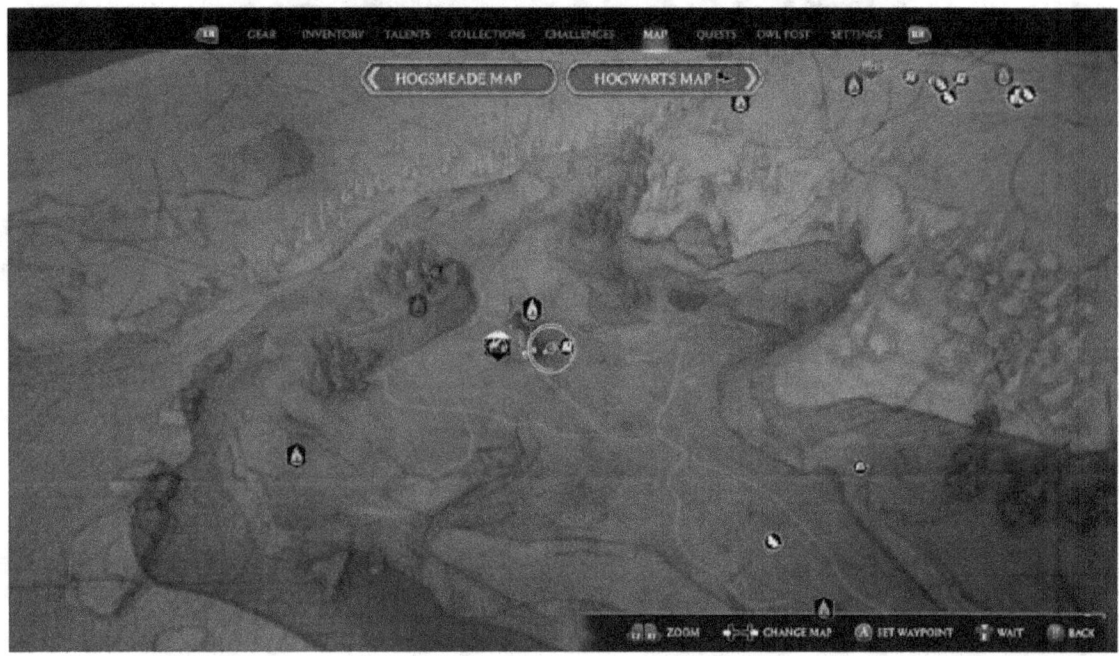

You can locate this Demiguise Statue at the main village of Feldcroft, located in the west central area of the Feldcroft region near the giant bandit camp castle that looms over the area. You'll need to give enough Moons to gain the level 2 Alohomora spell before you can reach this Demiguise, as it is locked away.

Upon arriving at the Feldcroft Village, travel east directly across from the Floo Flame and through the center courtyard to the middle house as you walk east. It's the one that lists two Collection Chests hidden away inside on your minimap.

Unlock the level 2 door with the improved Alohomora spell, and inside you'll find the Demiguise Location on a window sill to the left of the entrance, near the stairs leading up to the second floor.

Marunweem Lake Demiguise Statue Location - Marunweem Hamlet

For this Demiguise Location, you won't need the unlocking spell, but you will need to know where to look. This Demiguise can be found far in the south region of Marunweem, which will require you to progress past the narrow Coastal Cavern guarded by goblins that connects the South Sea Bog to the Poidsear Coast.

Upon reaching the fishing hamlet of Marunweem, head Northeast along the road from the Floo Flame until you reach a side quest giver named Marianne Moffett. Just to the left of her is a small house, on the east side of the small river running through the hamlet.

Enter the unlocked home to find some Collection Chests, as well as a Demiguise Statue that has been hidden behind a counter on the left side of the first floor of the house.

Manor Cape Demiguise Statue Location - Bainburgh

You can locate this Demiguise Statue at fishing hamlet called Bainbugh, located at the northern tip of the Manor Cape region, which is just a short hike south from the other hamlet in Marunweem. You'll need to give enough Moons to gain the level 2 Alohomora spell before you can reach this Demiguise, as it is locked away.

After arriving at the fishing hamlet, travel northeast from the Floo Flame while hugging the left side of the town, and go past the outdoor shopkeeper Agnes Coffey to slip past a bulletin board and find a tall house.

Using the Alohomora 2 spell to unlock the door, you can find the Demiguise Statue on the first floor as you enter, located on a small table next to one of the two armchairs in the house.

Cragcroftshire Demiguise Statue Location - Cragcroft

For this Demiguise Location, you won't need the unlocking spell, but you will need to know where to look. One of the last hamlets you can find in the open Highlands region, Cragcroft is located far in the southwest, over on the western corner of Cragcroftshire to the west of Manor Cape and below Marunweem Lake.

Upon arriving in town, head northeast from the Floo Flame building and cross the large tree in the middle of the village, to find a house just past the outdoor merchant named Bella Navarro.

Head past her stall and into the unlocked home, and climb the stairs on the left to reach the second floor of the home. Here you'll find the last Demiguise Statue on the far corner of the bedroom.

ALL MERLIN TRIALS SOLUTIONS

✧ South Hogwarts Region Merlin Trials

South Hogwarts Region Merlin Trial 1

- Spells Required: Incendio

Though you may find Merlin Trial altars if you explore during the early game, the first Merlin Trial cannot be solved until you complete the Main Quest, Trials of Merlin, in the South Hogwarts Hamlet of Lower Hogsfield. As you progress through Hogwarts Legacy's Main Story, you'll soon visit Lower Hogsfield to complete the Main Quest, The Girl from Uagadou. Immediately after this quest comes to a close, you'll hear Nora Treadwell in distress, so head over to the nearby marker to begin the Trials of Merlin quest.

Upon completing this quest, Nora will teach you about Merlin Trials and the rest of the trials in The Highlands will become discoverable.

To complete this Merlin Trial, use Incendio to light the braziers atop the nearby columns with the goal of having all three braziers lit at the same time. Once a brazier has been lit, its column will begin to descend into the ground so make sure to sprint from column to column to avoid having to start over.

South Hogwarts Region Merlin Trial 2

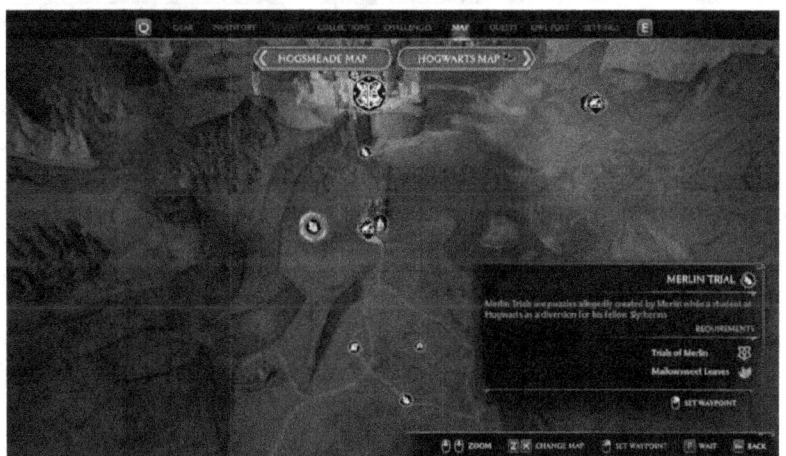

- Spells Required: Accio

Located just west of Lower Hogsfield, this Merlin Trial requires you to find three nearby altars and use the spell Accio to pull stone spheres onto them. To find the altars and the spheres you'll need to use the spell Revelio and search for objects highlighted in blue. Each altar will have an adjacent set of spheres so pull the spheres toward the altars and they'll automatically move into position.

Once all three altars have been filled by their corresponding spheres, you'll have successfully completed this trial.

- Altar 1 - Found just west of the Merlin Trial starting point.

- Altar 2 - Found down the slope northeast of the Merlin Trial starting point.

- Altar 3 - Found a few paces east of the Merlin Trial starting point.

South Hogwarts Region Merlin Trial 3

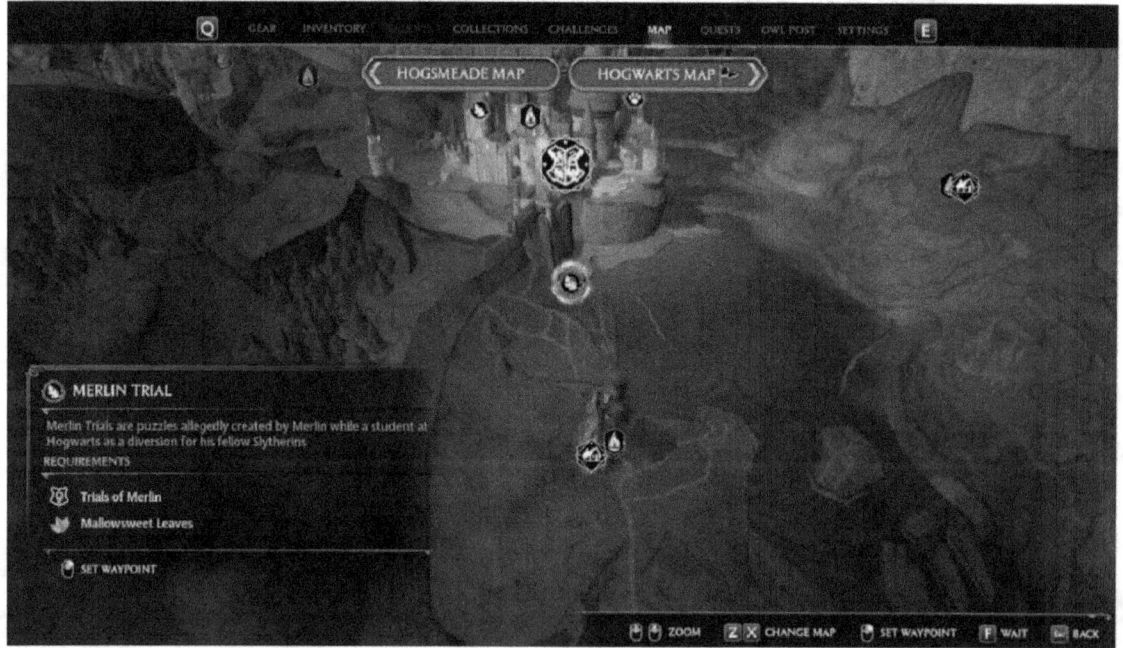

- Spells Required: Basic Cast

Found just south of Hogwarts' southern exit, scatter Mallosweet Leaves on the spherical Merlin altar to spawn a large stone orb to the southwest. Your objective is to roll the stone orb into the crater located on the southeastern shore. To get the orb down the hill and into the crater, stand facing the direction in which you want the orb to move and hit it with your Basic Cast four times in quick succession. The fourth strike in this combo will nudge the orb down the hill and into the crater, though, don't be surprised if you don't get a hole-in-one.

If the orb misses, simply head down to the shore and use another four-strike-combo to putt the orb into the hole.

South Hogwarts Region Merlin Trial 4

- Spells Required: Confringo

To reach this Merlin Trial, head east of Lower Hogsmeade and across the Great Lake. Once you've found the trial, sprinkle Mallowsweet Leaves and use the spell Confringo to destroy the five boulders marked with green carvings. Each of the five boulders can be seen from the Merlin Trial altar, so if you have trouble locating them, try standing on the altar and slowly rotating your POV.

South Hogwarts Region Merlin Trial 5

- Spells Required: Accio

Found south of the Hamlet of Aranshire, this Merlin Trial will task you with locating nearby stone orbs and using the spell Accio to bring the orbs to one of the three altars in the area. Though nearly identical to South Hogwarts Region Merlin Trial 2, these orbs will not automatically place themselves on the altars after you cast Accio. Instead, you must hold the spell and drag the orbs to an altar.

If you use Accio on the orbs and one or more of them get left behind, don't worry. As long as you drag a single orb onto on altar it'll register successfully.

- Altar 1 - South of the Merlin Trial starting point and its corresponding orbs are in the ridge to the north.

- Altar 2 - Located across the trail to the east and its orbs sit a few steps southeast of Altar 2.

- Altar 3 - This altar can be found directly north of the Merlin Trial start point. To locate the orbs, follow the trail north heading toward Aranshire and you'll find them sitting in front of the short stone wall.

South Hogwarts Region Merlin Trial 6

- Spells Required: Lumos

This Merlin Trial sits east of Merlin Trial 5 and southeast of the Hamlet of Aranshire. After spreading your Mallowsweet Leaves, butterflies will spawn directly in front of the Merlin altar. Use the spell Lumos to attract the butterflies to your wand, then escort them to one of the stone housings found down the cliffside. While the stone housings can be found using Revelio, the butterflies will not get marked by the spell. Thankfully, you'll always be able to see a kaleidoscope of butterflies from an empty stone, so make sure not to stray too far.

- Butterfly Stone 1 - After collecting the butterflies near the altar, head southwest of the Merlin Trial starting point to place them in the first stone.

- Butterfly Stone 2 - Found further southwest of Butterfly Stone 1 and the closest kaleidoscope of butterflies can be found flying northwest toward Hogwarts.

- Butterfly Stone 3 - Located near the small camp northwest of the trial starting point and you'll likely spot this stone while collecting the butterflies for Stone 2. To find the butterflies, stand near the stone and look directly north.

South Hogwarts Region Merlin Trial 7

- Spells Required: Basic Cast

After beginning this trial, look north and upward to find a large stone orb sitting on the cliffside. Make your way up to the orb and from here, you should spot a crater in the marsh to the south. Stand in front of the orb while facing the crater and use your Basic Cast four times in a row to hit the orb into the crater below.

South Hogwarts Region Merlin Trial 8

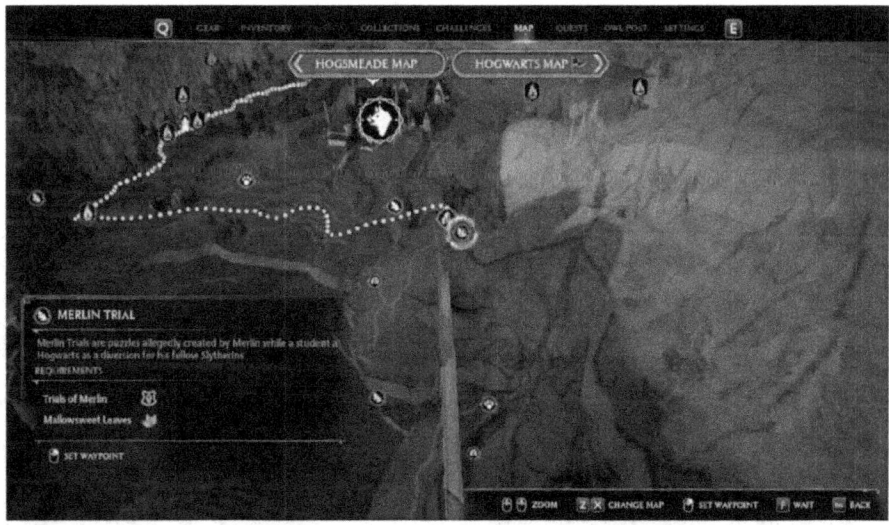

- Spells Required: Incendio or Confringo

Located just east of Hogsmeade Station, this trial is similar to the first one you completed during the Trials of Merlin main quest. Once you've started the trial, use Revelio to identify the location of three brazier-topped columns that you'll need to ignite quickly so all of them are lit simultaneously.

- Brazier 1 - Found north of the trial starting point toward Leopold Babcocke's vendor stall.

- Brazier 2 and 3 - Found southwest toward the train tracks.

However, unlike Merlin Trial 1, this trial has columns of different heights, meaning you'll need to prioritize igniting the taller columns first, or else the flames will extinguish before you can light all three braziers. Because Brazier 1 is the closest to the ground, use Incendio or Confringo on Braziers 2 and 3 before running north to light Brazier 1 last. If you've unlocked the spell Confringo, it's highly recommended that you use it to ignite the braziers instead of Incendio because Confringo's long-range will save you some time.

South Hogwarts Region Merlin Trial 9

- Spells Required: N/A

In one of the more straightforward Merlin Trials in South Hogwarts, this trial is less of a puzzle and more of a traversal challenge. Use Mallowsweet Leaves to begin the trial, then hop onto the westernmost stone and leap from stone to stone until you reach the final one. If you happen to fall off, restart the stone obstacle course until you make it from start to finish without touching the ground.

South Hogwarts Region Merlin Trial 10

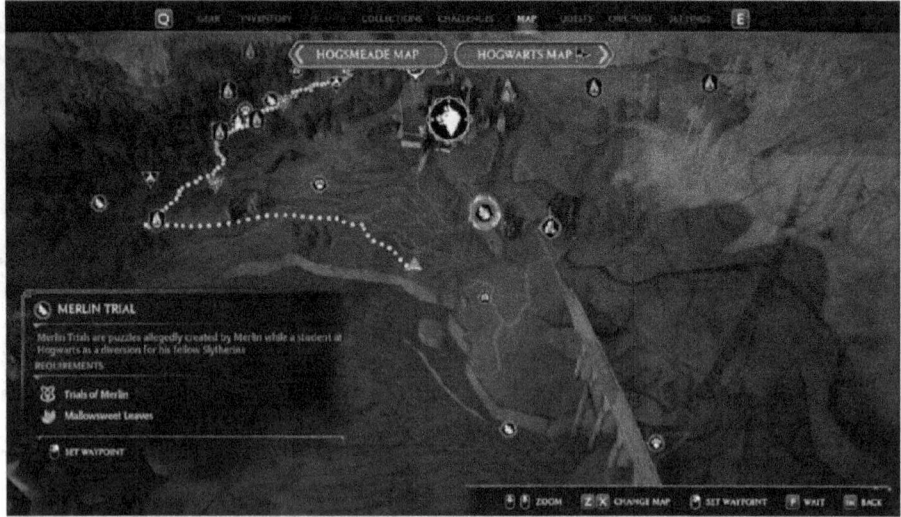

- Spells Required: Basic Cast

This Merlin Trial can be found just north of Hogsmeade Station and it'll task you with destroying nine orbs that surround the Merlin altar. Use Revelio to reveal the location of the orbs, and from the Merlin altar, you'll be able to spot and destroy eight of the nine orbs. To destroy the last orb, however, you'll need to head northwest toward the river and you should see it across the water. Your Basic Cast has limited range so, if necessary, walk closer to the orb before launching your attack. Confringo is also a viable substitute with longer range if you have it unlocked.

South Hogwarts Region Merlin Trial 11

- Spells Required: Lumos

Sat on the Great Lake's eastern shore, this Merlin Trial will again ask you to find three groups of butterflies so you can escort them to a stone housing. Once you've found a kaleidoscope of butterflies, use Lumos to attract them to your wand, then lead them to one of the three stones described below.

- Butterfly Stone 1 - Located east of the Merlin altar and its corresponding group of butterflies is even further east.

- Butterfly Stone 2 - Hidden behind the rock formation that is west of the Merlin altar and the stone's corresponding butterflies can be found hovering above the Great Lake.

- Butterfly Stone 3 - This stone is north of the Merlin altar and its butterflies can be found just past the stone on the shore.

South Hogwarts Region Merlin Trial 12

■ Spells Required: Basic Cast

To find this Merlin Trial, you'll need to travel to the western reaches of South Hogwarts to the area past the Quidditch Pitch. After sprinkling Mallowsweet, look to the top of the rock formation to the northeast to spot a large stone orb. Carefully climb or fly up to the top of the rock formation and peek off the northern edge to locate the crater this orb must fill.

Now all that's left to do is to strike the orb with your Basic Cast four times in quick succession to hit the orb toward the crater. This will likely take several attempts so follow the orb down the hill and nudge it into the crater to complete this trial.

A chest containing Gear can be found near the rock west of the Merlin altar.

South Hogwarts Region Merlin Trial 13

■ Spells Required: Lumos

This Merlin Trial can be found by following the river west of the Owlery until you reach the waterfall on the western reaches of South Hogwarts. This trial is another butterfly challenge so you'll need to find three sets of butterflies and escort them to a stone housing by using the spell Lumos. Below we'll provide a list of exactly where the stones and butterflies can be found:

■ Butterfly Stone 1 - This stone is rather close to the Merlin altar as it's just a few steps to the east. The closest group of butterflies can be found southeast before the stone bridge.

■ Butterfly Stone 2 - Found on the ledge west of the Merlin altar, while the stone's corresponding butterflies are located to the northwest. To escort the butterflies to the stone you'll need to mantle the ledge which will stop the insects from following your wand. After mantling, simply cast Lumos again to bring them up to the stone.

■ Butterfly Stone 3 - To reach this stone, follow the trail that leads southeast, cross the bridge, and work your way up the cliff to the right. Upon reaching the stone, you should be able to see the final butterflies you'll need to the northwest.

South Hogwarts Region Merlin Trial 14

- Spells Required: Confringo

To avoid having to climb steep cliffs, it's recommended that you head toward this Merlin Trial by jogging up the hill west of the Hamlet of Lower Hogsfield or, if possible, using your broom to fly up to the area. Once you've reached this trial that overlooks Hogwarts from the southwest, you'll notice that several stones in the surrounding area have been marked with green carvings. Use Confringo to destroy all five of the stones and you can check this trial off your list.

South Hogwarts Region Merlin Trial 15

- Spells Required: Accio

Located east of Aranshire this Merlin Trial will require you to play another round of wizard golf and move a large stone orb into a large crater. Once you've started the trial, look down the southern cliff and in addition to seeing the crater you need to fill, you should spot a path climbing up the hill to the south. Follow the path up the hill and you'll find a well to your right. To the right of the well sits the stone orb you'll need.

This well is the starting point for the Side Quest, "Well, Well, Well".

Unlike similar trials you've completed thus far, this stone orb has a purple glow to indicate that it can be easily manipulated by the spell Accio. Use Accio to drag the orb down to the crater and congrats! You have now completed all 15 Merlin Trials in the South Hogwarts Region!

✧ Hogsmeade Valley Merlin Trials

Hogsmeade Valley Merlin Trial 1

NOTE: This trial requires Lumos.

The first Merlin Trial is located east of Hogsmeade, near the edge of the mountains. Activate it, then you'll need to clear out the area before you can get to work.

Head east, down the little path to the main road and clear out all the poachers, where you'll find the first group of glowing butterflies. Cast Lumos and walk into them, and they'll gather around your wand, then continue up the road and use basic wand blasts to destroy the wooden fence to find a wolf.

Defeat it, then approach the second group of butterflies with Lumos, and they'll join you, too. Turn right and climb up the small ledges to find the third butterfly group. Cast Lumos again and they'll join you, too. If any groups of butterflies fall behind, grab them with Lumos and then approach all three stone plinths around the trial starting area, and each butterfly group will go inside one, completing the trial.

Hogsmeade Valley Merlin Trial 2

The second Merlin Trial is found near Falbarton Castle, on the northeastern side of Hogsmeade Valley.

From the castle gate, turn left and hop into the creek. Run straight ahead and cast a fire spell on the vines to clear them, then cast Accio on the grate to pull it loose and create a gap you can crawl through.

Head left and climb the ledges to reach the starting point. After activating it, hop back down into the tunnel to get the first butterfly swarm and use Lumos to bring it with you as you climb back up, then gather the two other butterfly swarms nearby.

Turn left and approach one of the stone plinths to deposit a butterfly swarm, then go up the hill to find the other two to complete the trial.

Hogsmeade Valley Merlin Trial 3

NOTE: This trial requires Repulso or Confringo.

The third Merlin Trial is found on the northeastern side of Hogsmeade Valley, between the East Hogsmeade Valley and Falbarton Castle fast travel points.

After activating the trial, head further up the hill to find a large, round stone. Just to the left of the stone is a long, slippery slope. Cast Repulso or Confringo on the rock to push it into the trough (any spell that moves the stone away from you will work), then wait until the stone falls into the giant stone hole at the bottom of the hill to complete the trial.

Hogsmeade Valley Merlin Trial 4

NOTE: This trial requires Confringo.

The fourth Merlin Trial is found on the northeastern side of Hogsmeade Valley, northwest of the East Hogsmeade Valley fast travel point.

Climb the small cliff to find the starting point, then activate the trial and climb up the nearby cliffs to get a good angle on the three braziers, then quickly shoot each of them with Confringo to complete the trial.

Hogsmeade Valley Merlin Trial 5

The fifth Merlin Trial is found just north of Hogsmeade, itself.

Take out the enemies in the area, then activate the trial. Use Revelio to highlight the various large stones surrounding the area, then destroy them all with spells like Confringo to complete the trial.

✦ North Hogwarts Region Merlin Trials

North Hogwarts Region Merlin Trial 1

The first Merlin Trial is just west of the Forbidden Forest fast travel point. Throw some Mallowsweet on the point and then turn west and use basic attacks to destroy all the small balls atop the pillars to complete the trial.

North Hogwarts Region Merlin Trial 2

NOTE: This trial requires the Reparo spell.

The first Merlin Trial is west of the East North Hogwarts Region fast travel point. It can be found along the river, near a treasure cave and a butterfly point.

After activating it with Mallowsweet, use Reparo on the four crumbled statues surrounding the trial start point to complete it.

North Hogwarts Region Merlin Trial 3

NOTE: This trial requires the Confringo spell.

The third Merlin Trial can be found on the far western side of the North Hogwarts Region.

There is a small tower to the northwest of the trial location, so cast Confringo on the vines across the door to reveal a brazier.

Cast Confringo on the brazier in the tower, then turn around and cast it on the one to the south. After that, run across the bridge and turn left, then cast Confringo on the brazier just up the hill to complete the trial.

North Hogwarts Region Merlin Trial 4

NOTE: This trial requires Lumos.

The fourth Merlin Trial requires the spell Lumos, and it can be found west of Hogsmeade. After activating it with Mallowsweet, run to one of the glowing groups of butterflies. There is one by the tent, one by the tree, and one by the cliffs.

Cast Lumos and they'll bunch around your wand, then lead them back to one of the stone structures surrounding the trial's starting point. Once you've brought each butterfly group back to a stone, the challenge is complete.

North Hogwarts Region Merlin Trial 5

NOTE: This trial requires Flipendo.

The fifth Merlin Trial can be found in the northern-most part of the North Hogwarts Region. After activating it with Mallowsweet, you'll need to cast Flipendo on the stone pillars.

One is right next to the fire, the next is across the bridge and down on the shore, then the third is on the small island in the middle of the pond. Activate them all to complete the trial.

✧ Forbidden Forest Merlin Trials

Forbidden Forest Merlin Trial 1

NOTE: This trial requires Confringo

The first Merlin Trial in the Forbidden Forest can be found just outside of Jackdaw's Tomb. Activate the trial and three braziers will appear. One is located to the south, one to the north, and one to the northwest near the lake.

First, burn the spiderweb near the southern brazier, then light it. Once it's lit, run to the north and light the next one, then stand next to it and look towards the lake to find the third. Once all three are lit, you're done!

Forbidden Forest Merlin Trial 2

NOTE: This trial requires Lumos, Confringo/Incendio

The second Merlin Trial in the Forbidden Forest can be found in the central region, east of the West Forbidden Forest fast travel point.

Clear out the poachers here, then activate the trial. Use Confringo on the stacked boulders to the west, which block the way to a cave holding the butterflies. Use Lumos to gather them around your wand and have them follow you.

Use Confringo again on the vines and tree branches blocking another one to the east, then gather the third and final butterfly swarm, then return them to the three glowing stone plinths surrounding the trial to complete it.

Forbidden Forest Merlin Trial 3

The third Merlin Trial in the Forbidden Forest can be found in the northeastern corner, just north of the North Ford Bog Entrance fast travel point.

Once you arrive, activate the trial, then shoot the cairn stone piles off the stone pillars, then use the wooden box to hop on top of the one nearest you.

Hop from pillar to pillar (or walk/run from pillars that are close together) without touching the ground. Once you've jumped across them all, you'll complete the trial.

✧ North Ford Bog Merlin Trials

North Ford Bog Merlin Trial 1

NOTE: This requires Accio, Confrigo/Incendio

This trial is in the southeastern section of North Ford Bog, very close to the East North Ford Bog fast travel point.

Activate the trial, then head southwest and you'll find some stone pillars surrounding a grassy area. Use Incendio/Confrigo on the grassy area to burn it away and reveal a stone hole.

Use Accio on the large, round boulder on the hill above the hole and pull it in to complete the trial.

North Ford Bog Merlin Trial 2

NOTE: This trial requires Flipendo

This Merlin Trial can be found east of San Bakar's Tower fast travel point, in northern North Ford Bog.

Clear out the Inferi, then activate it and Flipendo stones will appear on the west, southeast, and east. Cast Flipendo on them and you'll complete the trial.

North Ford Bog Merlin Trial 3

NOTE: This trial requires Confringo

This Merlin Trial can be found just east of North Ford Bog fast travel point, in northern North Ford Bog.

Activate the trial and you'll need to use Confringo on the many statues along the cliffs over the water. Several are visible right away, but you'll need to use Confringo on the grassy vines hanging from the cliffs to expose the rest.

Use Revelio if you have trouble seeing them, as they'll glow blue. Just destroy them all to finish the trial.

North Ford Bog Merlin Trial 4

This Merlin Trial can be found in Pitt-Upon-Ford, near the fast travel point.

Activate the trial and then climb atop the stone platforms nearby. Jump from stone platform to platform without touching the ground as you ascend.

Jump all the way to the top, touching only the stone platforms, to complete the trial.

✧ Feldcroft Region Merlin Trials

Feldcroft Region Merlin Trial 1

■ Spells Required: Reparo

To find this Merlin Trial, start at either Rockwood Castle or the Feldcroft Hamlet before following the trail heading north. This Merlin Trial will be on the trail but be wary of the various enemies that typically patrol this area. Once you've activated the Merlin Trial, use Revelio to reveal the location of three destroyed statues. Use Reparo to fix all three statues and you'll complete this Merlin Trial.

- Statue 1 - Found northwest of the Merlin Trial starting point, just across the dirt path.

- Statue 2 - Found atop the eastern cliff above the Merlin Trial starting point.

- Statue 3 - Found further along the trail heading northward.

Feldcroft Region Merlin Trial 2

- Spells Required: Basic Cast

Found north of Rockwood Castle on the cliffs facing the miniature lake, this Merlin Trial will require you to destroy nine orbs that are grouped in threes. After beginning the trial, look to the southeast and destroy the three orbs that get marked in blue when you use Revelio. Next, look southwest and destroy the second trio of orbs on the hill above.

The last set of orbs, however, is across the lake to the northwest as seen in the image above. Either make the trek around or fly to the other side to get in range of your Basic Cast. Alternatively, if you walk up the southwestern hillside you can get close enough to use the spell Confringo.

❖ Feldcroft Region Merlin Trial 3

- Spells Required: Confringo

This Merlin Trial is located northwest of Rockwood Castle and it overlooks the river that wraps around the western Feldcroft Region. Start by scanning the area with Revelio to reveal the location of three nearby braziers. Your goal is to quickly light the braziers so all three of them are ablaze simultaneously.

To accomplish this, climb up the southeastern cliff and light the westernmost brazier before running northeast to light the second brazier. From here, look down the cliff toward the north and light the brazier below to complete this Merlin Trial.

If you find it difficult to get all three braziers lit at the same time, remember to maximize Confringo's range to minimize the amount of time you waste running from brazier to brazier.

Feldcroft Region Merlin Trial 4

- Spells Required: N/A

Once you've arrived at this Merlin Trial west of Rockwood Castle, sprinkle Mallowsweet Leaves on the altar and look toward the water to find four stone platforms. To complete this trial, hop onto the rightmost platform and jump across the stones until you reach the fourth and final one.

You must make it across all four platforms in a single attempt to check this Merlin Trial off your list. If you happen to miss a jump and go for a swim, simply head back to the longest platform and try again.

Feldcroft Region Merlin Trial 5

- Spells Required: Confringo

To reach this Merlin Trial, head to the South Feldcroft Floo Flame and travel due west until you hit the nearby coast. Once you've activated the Merlin Trial, use Revelio to mark five stones in the area that have distinct green-colored carvings. Blast all five stones with Confringo to destroy them and complete this trial.

- Stone 1 - Found just southwest of the Merlin altar.

- Stone 2 - Found atop the rock formation to the northeast.

- Stone 3 - Found on the southern coast.

- Stone 4 - Found atop the rock formation south of the Merlin altar.

- Stone 5 - Found atop the eastern rock formation.

Feldcroft Region Merlin Trial 6

- Spells Required: Basic Cast, Depulso, or Accio

After you activate this Merlin Trial that's found on the coast south of the Hamlet of Feldcroft, turn around to face the northeast and you'll find a large crater in the ridge ahead. To find the orb that you'll need to complete this trial, travel northeast to reach the end of the ridge and the orb can be found on a cliff to your left.

Now you'll need to escort the orb into the crater at the bottom of the ridge. This can be done by hitting the orb with a four-hit combo of your Basic Cast or by using Depulso or Accio to maneuver the orb down the hill and into the crater.

Feldcroft Region Merlin Trial 7

- Spells Required: Confringo

The seventh Merlin Trial sits just southwest of the North Feldcroft Floo Flame and on the outskirts of the nearby Small Bandit Camp. Because of its proximity to the Bandit Camp, you should prepare for a battle as you may be spotted by patrolling Ashwinder Soldiers.

After spreading Mallowsweet and activating the trial, use Revelio to locate five stones in the area that are marked with green carvings. Use Confringo to destroy the stones described below and you'll wrap up this Merlin Trial.

- Stones 1 & 2 - Found east of the Merlin altar.

- Stone 3 - Located on the rock northeast of the Merlin altar.

- Stones 4 & 5 - These stones sit on the hillside to the southwest, but they will be blocked from your view by the nearby structure if you don't use Revelio to mark their location.

Feldcroft Region Merlin Trial 8

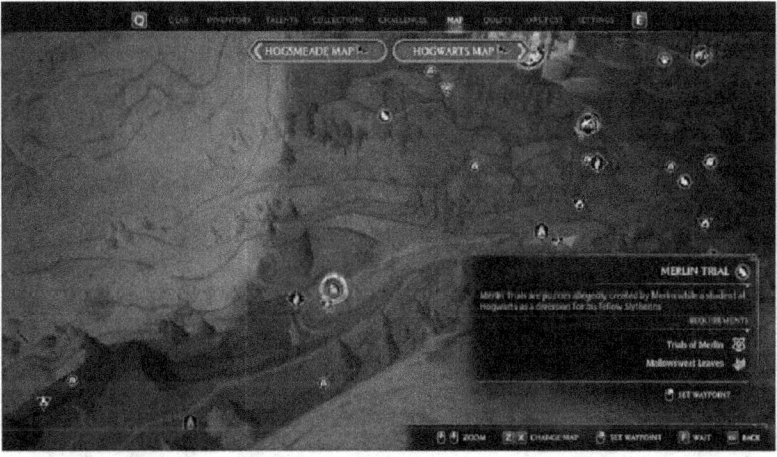

- Spells Required: Flipendo

This Merlin Trial is located toward the northern end of Feldcroft near the Small Bandit Camp that you may have visited as part of The Lost Child Natty Quest. Found just east of the Ancient Magic Hotspot in the area, this Merlin Trial tasks you with rotating stone cubes to match the symbols on the cubes below them.

When using Flipendo to rotate the cubes. a single cast of Flipendo will rotate the cube in the direction you're facing one time. However, it's not enough to simply get the correct symbols on the same plane, you also need to ensure that the symbols aren't upside down or turned at an incorrect angle.

When you get the cube symbols lined up perfectly, they'll get covered in vines to indicate that you can move on to the next set of cubes. To finish this Merlin Trial, you'll need to solve all three of the cube puzzles surrounding the encampment.

Feldcroft Region Merlin Trial 9

- Spells Required: N/A

This Merlin Trial can be found by traveling to the North Feldcroft Floo Flame before heading southwest into the closest Small Bandit Camp. Because this trial is found at the center of the Small Bandit Camp, you won't have the luxury of avoiding combat. Take down Ranrok's Loyalists and you can then peacefully spread your Mallowsweet to start the trial.

To complete the Merlin Trial, climb atop the westernmost stone block and hurdle over to the following blocks. When you reach the third block your path may be blocked by wooden crates and barrels. Destroy them with the spell of your choosing then leap to the end of this makeshift obstacle course to wrap up this trial.

Feldcroft Region Merlin Trial 10

- Spells Required: Flipendo

Found southwest of the twin Bandit Camps in North Feldcroft, this Merlin Trial will task you with destroying five stones marked with green carvings. Though four of these stones can be seen from the Merlin Trial altar, the fifth one is slightly trickier to find. If you're standing on the Merlin altar, your view of this stone will be obstructed by the southern rock formation.

Head down the path leading southwest and look to your left to find and destroy the stone.

Feldcroft Region Merlin Trial 11

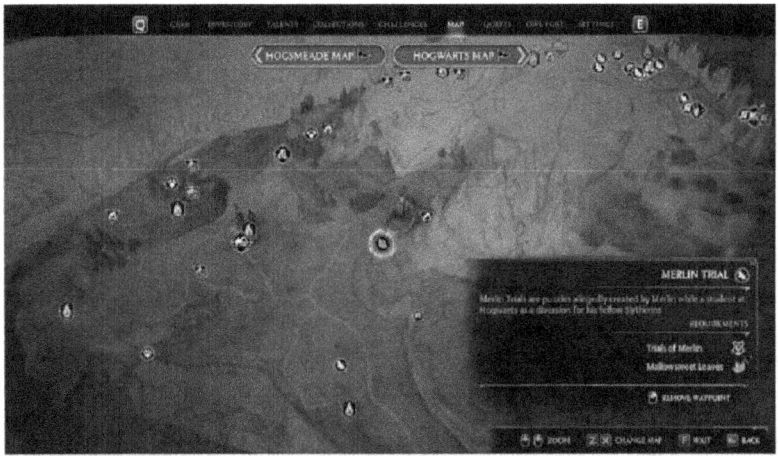

- Spells Required: Basic Cast

To discover this Merlin Trial, head east of the Feldcroft Hamlet toward the Troll Lair and Treasure Vault. After sprinkling Mallowsweet on the altar, search for and destroy the nine orbs in the surrounding area. The orbs will be split into groups of three and they're found in the areas described below.

- Orb Trio 1 - Found just northwest of the Merlin altar.

- Orb Trio 2 & 3 - The remaining six orbs are in the ruins northeast of the Merlin altar.

Feldcroft Region Merlin Trial 12

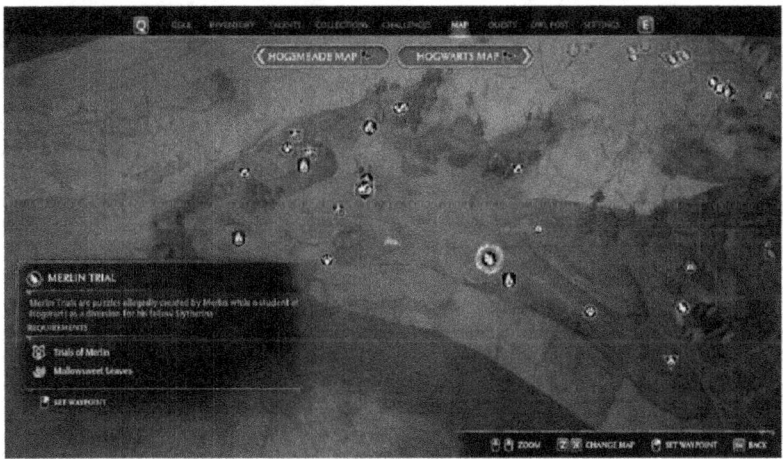

- Spells Required: Confringo

The twelfth Merlin Trial in Feldcroft sits just northwest of the Feldcroft Catacomb Floo Flame. To complete this trial you'll need to use the spell Confringo to destroy five stones in the area that are marked with green carvings. If you have trouble locating the stones, remember that you can use Revelio to reveal their location.

Though this may seem relatively straightforward if you've completed the earlier Merlin Trials listed in this guide, this trial has a bit of a surprise.

One of the green-carved stones can be found at the base of a large rock formation to the west of the Merlin Trial start point. Once this stone it's destroyed, you'll reveal a secret passage covered in some foliage. Use Confringo or Incendio to burn down the foliage and you'll be rewarded with some Gear from a special chest.

Feldcroft Region Merlin Trial 13

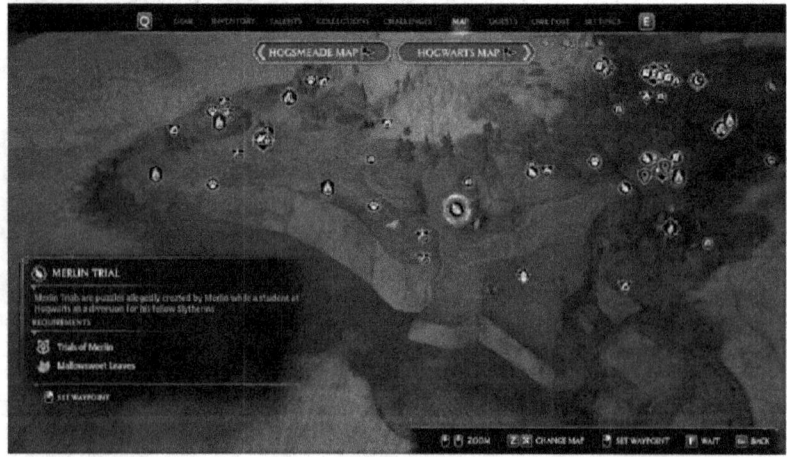

- Spells Required: Lumos

This butterfly-related Merlin Trial is located between the Feldcroft Catacomb Floo Flame and the Feldcroft Battle Arena. Start the trial by sprinkling Mallowsweet Leaves then use Revelio to ping the nearby stone enclosures. While the stone enclosures are all within a few feet of the Merlin Trial altar, the butterflies are quite a bit trickier to find.

Once you've found the butterflies described below, use the spell Lumos to guide them to any of the three stone enclosures surrounding the Merlin Altar.

- Butterflies 1 - Found below the rock formations south of the Merlin altar (pictured above).

- Butterflies 2 - Follow the path heading east and go up the steps to find the butterflies at the top of the hill

- Butterflies 3 - From the Merlin altar look atop the western cliffs to spot the final group of butterflies.

Feldcroft Region Merlin Trial 14

- Spells Required: Confringo

This Merlin Trial can be found almost perfectly between the Feldcroft Catacomb Floo Flame and the Hamlet of Irondale. After spreading Mallowsweet to begin the trial, use Revelio to reveal five nearby stones and blast them with your Confringo spell. The five stones are all visible from the Merlin Trial start point.

Because of this trial's proximity to a Small Bandit Camp, be prepared to take down the nearby Loyalists that will likely interrupt you.

Feldcroft Region Merlin Trial 15

- Spells Required: Confringo

To reach this Merlin Trial, head to the Hamlet of Irondale and follow the path heading southwest to cross the small stone bridge. To successfully complete this trial you'll need to light three nearby braziers in quick succession so all three of them are ablaze simultaneously.

Because your goal is to have all three braziers lit at the same time, start by using Confringo on Brazier 2 first before running back across the bridge to light Brazier 3. Brazier 1 is the shortest of the three so it should be lit last or you'll likely run out of time.

- Brazier 1 - Found at the edge of the waterfall east of the Merlin altar.

- Brazier 2 - Located across the bridge to the northeast.

- Brazier 3 - Just southwest of the Merlin altar.

Feldcroft Region Merlin Trial 16

- Spells Required: Basic Cast

The final Merlin Trial in the Feldcroft Region is found in the center of the Irondale Hamlet. Once you've started the trial, use Revelio to reveal the location of nine orbs in the surrounding area and destroy them with the spell of your choosing. As always, the orbs will be grouped into trios so check the list below to learn their exact locations.

- Orb Trio 1 - East of the Merlin altar just in front of the cottage.

- Orb Trio 2 - Located atop the hillside northwest of the Merlin altar.

- Orb Trio 3 - Found down the southern hill near the Irondale Floo Flame.

✧ Marunweem Lake Merlin Trials

❖ Marunweem Lake Merlin Trial 1

- Spells Required: Basic Cast

This Merlin Trial can be found by traveling to the Marunweem Floo Flame before heading west. The Merlin altar will sit just in front of some stone structures and a dock. Activate the Merlin Trial by sprinkling Mallowsweet and then walk onto the stone walkway directly in front of the Merlin altar.

Jump across to the following stones, and use Revelio to reveal the area of the dock you'll need to land on. Landing on any non-highlighted areas will force you to restart this obstacle course. Now that you're at the center of the dock, leap onto the tilted stones, slide to the end of the path, and shoot the nearby wooden barrel

Landing on the barrel will reset your obstacle course run. Be sure to clear it before attempting to leap to the following platform.

With the wooden barrel out of the way, you can safely jump onto the small stone platform before leaping once more to reach the final platform that has a Bag of loot on it.

Marunweem Lake Merlin Trial 2

■ Spells Required: Flipendo

Found toward the eastern reaches of the Marunweem Lake region, this Merlin Trial will task you with rotating three cubes until it perfectly matches the symbols beneath them. Use Revelio to mark the cubes in the surrounding area and once you've found them, use Flipendo to rotate the cubes to match the symbols. Remember that the symbols must match and the symbols must be facing the same direction for you to solve each puzzle.

When you've successfully lined up the symbols, the stone cubes will get covered in vines to indicate that you can move on to the next set of cubes. Read the list below to learn the location of the three cubes:

■ Cube 1 - Southeast just behind the Merlin altar.

■ Cube 2 - Northeast of the Merlin altar, roughly where the Thestral Den map marker is located.

■ Cube 3 - Found up on the cliffside southwest of the Merlin altar near an Astronomy Table.

Marunweem Lake Merlin Trial 3

- Spells Required: Depulso and Wingardium Leviosa

Located on the hillside east of the Marunweem Lake Floo Flame, this Merlin Trial will ask you to open a gate to free a stone orb that will later need to be placed into a crater. To open the gate, use Depulso on the mechanism pictured above. With the gate open, climb the platform ahead to locate the stone orb and cast Wingardium Leviosa on it to carry the orb with you.

With the orb in tow, walk across the small bridge heading east, then turn left to drop the orb into its corresponding crater.

Marunweem Lake Merlin Trial 4

- Spells Required: Lumos

Located just northwest of the Marunweem Ruins Floo Flame, this Merlin Trial will require that you use Lumos to lure butterflies into the three stone housings in the area. To reveal the location of the stone housings needed for this trial, use Revelio and they'll be marked in blue. The butterflies are more elusive and they will not be marked by your spell so after sprinkling Mallowsweet Leaves, refer to the list below to learn where you can find the three groups of butterflies:

- Butterflies 1 - Floating above the campfire northeast of the Merlin altar.

- Butterflies 2 - Beside barrels just north of the Merlin altar, as seen in the image above.

■ Butterflies 3 - These butterflies are found southeast of the ruins floating around a pointing statue.

✧ South Sea Bog Merlin Trials

South Sea Bog Merlin Trial 1

■ Spells Required: Lumos

Located toward the northern area of the South Sea Bog, this Merlin Trial will task you with gathering butterflies and guiding them to a stone enclosure with the spell Lumos. Though butterflies cannot be pinged by Revelio, all of the butterflies in this trial can be seen from one of the three stone enclosures.

Once you've found a kaleidoscope of butterflies, approach them and activate Lumos to lure them to an enclosure. Deactivating Lumos or casting another spell will make the butterflies flee.

■ Butterflies 1 - Found north of the Merlin Trial altar, to the right of a tree.

■ Butterflies 2 - Found southwest of the Merlin Trial altar at the base of another tree.

■ Butterflies 3 - Found southeast of the Merlin Trial altar but they can be tricky to see behind all of the branches in your way.

South Sea Bog Merlin Trial 2

- Spells Required: Basic Cast

To get to the final Merlin Trial in South Sea Bog, head to the southern reaches of the region just past the Medium Bandit Camp. Though you may elect to clear the camp before taking on the Merlin Trial, this is not required as the Trial start point is on a cliffside devoid of enemies.

Spread your Mallowsweet Leaves to begin the Merlin Trial, then turn to the southeast where you'll spot three of the nine orbs you'll need to destroy. The second trio of orbs is found under the cliff to the north and the third trio is set northwest near the small body of water.

All of the nine orbs can be destroyed from the cliffside if you want to avoid battle.

✧ Manor Cape Merlin Trials

Manor Cape Merlin Trial 1

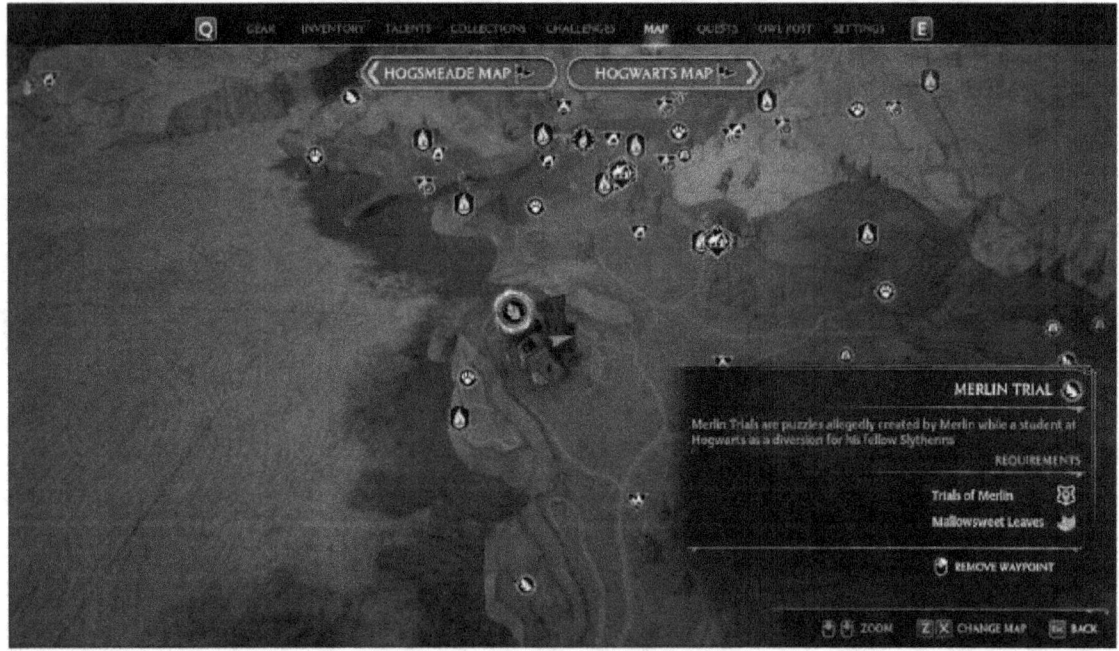

- Spells Required: Confringo or Incendio

This Merlin Trial is found toward the northwestern end of the West Manor grounds near a derelict fountain. To complete the trial, you'll need to locate three braziers in the surrounding area and light them so all three of them are ablaze at the same time. Because the braziers are atop pillars of varying heights, it's best to light the braziers in the optimal order to ensure that you have time to light all three braziers. Follow the order listed below to optimize your chances of keeping the braziers lit simultaneously:

- Brazier 1 - Near the gazebo on the hill to the west of the Merlin altar.

- Brazier 2 - At the top of the small set of stairs facing the hedge maze that sits south of the Merlin altar.

- Brazier 3 - Located directly in front of the altar to the northeast.

Manor Cape Merlin Trial 2

- Spells Required: Flipendo

The nearby, fan-like rotating mechanisms are not associated with this Merlin Trial. They're linked to the Treasure Vault just east of the Merlin altar.

Located east of the West Manor Cape Floo Flame, this Merlin Trial will task you with rotating three cubes in the surrounding area until the symbols on the cubes match up with their corresponding pedestals. To rotate the cubes, cast Flipendo while facing the direction you want the cube to rotate. When the symbols match up perfectly, the cubes will get covered in vines to indicate that you can move on to the next cube. Read the list below to learn where you can find the three cubes needed to complete this Merlin Trial:

- Cube 1 - North of the Merlin altar, near one of the fan-like rotating mechanisms.

- Cube 2 - Atop the southwestern cliff, just above a pointing statue.

- Cube 3 - On the shore east of the Merlin altar.

Manor Cape Merlin Trial 3

- Spells Required: Depulso or Wingardium Leviosa

To reach this Merlin Trial, travel to the West Manor Cape Floo Flame before heading south toward the ruins near the Hippogriff Den. You'll find the Merlin altar at the bottom of a steep hill on the shore. After activating the Merlin Trial, approach the large stone orb west of the Merlin altar. Your goal is to get this orb up the steep northern hill and into the crater at the top. To accomplish this, you have the following options:

- Depulso - The first way to get the orb up the hill is to spam the orb with the spell Depulso to nudge it up the steep hill. This can be tricky, but if you're quick on the draw this option will save you some time.

- Wingardium Leviosa - Alternatively, you can cast Wingardium Leviosa on the orb and escort it up the path leading to the top of the eastern hillside. When you reach the area pictured above, drop the orb so it can fall into the crater below.

Manor Cape Merlin Trial 4

- Spells Required: Lumos

This Merlin Trial is found on the southern half of Manor Cape, just north of the arching bridge. To complete this trial, you'll need to find three sets of butterflies and use the spell Lumos to lure the butterflies into stone enclosures surrounding the Merlin altar. Though the stone enclosures can be found by using Revelio, the butterflies are more elusive. Refer to the list below to learn where to find the three sets of butterflies:

- Butterflies 1 - By the lantern that leads to the nearby bridge.

- Butterflies 2 - Found beneath the first archway of the nearby bridge.

- Butterflies 3 - From the Merlin altar, look southeast and drop to the platform below. Turn 180 degrees to face the Merlin altar and you'll notice some black bramble. Activate Lumos to walk past the bramble and you'll find the butterflies. Escort them up to the final stone housing to complete this Merlin Trial.

Manor Cape Merlin Trial 5

- Spells Required: Confringo

The final Merlin Trial in Manor Cape is found near the ruins on the southeastern tip of the peninsula. To complete the trial, you'll need to locate five stones marked with green carvings. Once you've found the stones destroy them by shooting them with Confringo.

✧ Cragcroftshire Merlin Trials

Cragcroftshire Merlin Trial 1

- Spells Required: Confringo

This Merlin Trial sits toward the western end of Cragcroftshire, just west of Rohan Prakash's Wanderer Shop and near the sole Ancient Magic Hotspot in this region. When you reach the Merlin Trial, defeat the Ashwinders patrolling the area, then sprinkle Mallowsweet on the altar to begin the trial. To complete this trial, you'll need to destroy five stones in the surrounding area with the spell Confringo. The stones can be identified by their green markings and they're found in the locations described below:

- Stone 1 - Behind the wooden wall northeast of the Merlin altar.

- Stone 2 - On the hill northwest of the Merlin altar, just past a decrepit tree.

- Stone 3 - Found east of the Merlin altar in front of a stone doorway.

- Stone 4 - Southeast of the altar in front of a tree with foliage.

- Stone 5 - The fifth stone is a bit trickier to find. From the altar, head southeast and search for a triangle-shaped hole in the ruins. As pictured above, the final stone can be seen through this triangle-shaped hole.

Cragcroftshire Merlin Trial 2

- Spells Required: Lumos

To reach this Merlin Trial, start at the Cragcroft Hamlet and head northwest toward the Spider and Mongrel Lairs. After spreading Mallowsweet Leaves onto the altar, use Revelio to reveal the location of three stone housings. You'll need to find three sets of butterflies and guide them to the stone housing by luring them with your Lumos spell. Check the list below to learn where to find the three sets of butterflies:

- Butterflies 1 - Above the campfire inches away from the Merlin altar.

- Butterflies 2 - Floating around the cart southeast of the Merlin altar.

- Butterflies 3 - Found in front of the wall directly northeast of the Merlin altar.

Cragcroftshire Merlin Trial 3

■ Spells Required: Confringo or Incendio

This Merlin Trial is located at the center of the Cragcroft Hamlet. After activating the trial, walk onto the path to the southwest and use Revelio to reveal the location of three braziers. To successfully complete this trial, you'll need to quickly light the three braziers so they're all ablaze simultaneously. Because the braziers sit atop pillars of varying sizes, you'll want to use Confringo or Incendio to light the brazier closest to the Merlin Trial first, before running down the path to the south to light the second and third braziers.

Cragcroftshire Merlin Trial 4

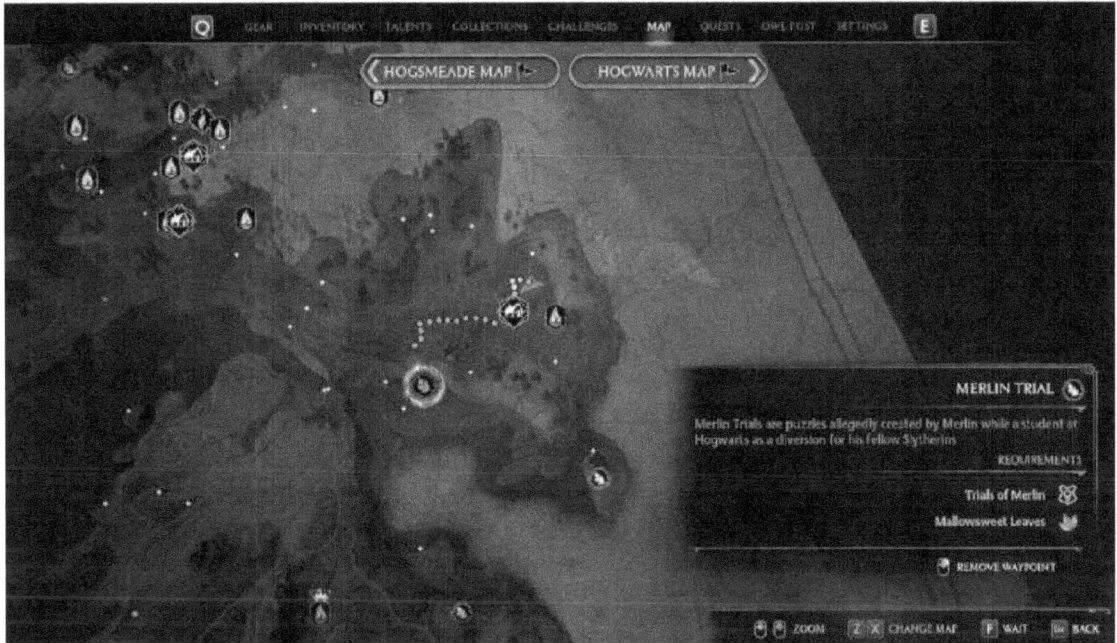

■ Spells Required: Basic Cast

Found near the shore southwest of the Cragcroft Hamlet, this Merlin Trial will task you with destroying nine orbs in the surrounding area. To reveal the orbs, cast the spell Revelio and scour the area for orbs marked in blue. The orbs can be found in sets of three and their exact locations are detailed in the list below:

■ Orb Trio 1 - Directly west of the Merlin altar.

■ Orb Trio 2 - High on the cliffside northeast of the Merlin altar.

■ Orb Trio 3 - On the shore to the southeast.

Cragcroftshire Merlin Trial 5

■ Spells Required: N/A

The fifth Merlin Trial in Cragcroftshire is found in the southeastern reaches of the region. Once you've started the trial, jump onto the rock formation northwest of the altar and cast Revelio to reveal Merlin's version of an obstacle course. Leap across the six stones in the obstacle course without touching the ground and you'll have completed the final Merlin Trial in Cragcroftshire.

ALL ARITHMANCY PUZZLE DOOR LOCATIONS AND SOLUTIONS

Need help solving those mysterious puzzle doors that are surrounded by beast symbols in Hogwarts Castle? This puzzle door guide will provide all the locations and solutions that you'll need to unlock the Arithmancy Doors in Hogwarts Legacy.

With a total of 11 Arithmancy Doors to discover and unlock, we've included a handy checklist, as well as all the potential loot types that you can expect from unlocking each door.

✧ Ravenclaw Tower

The Ravenclaw tower may be the first door you come across if you align yourself with the best house. It can be found right under the staircase next to a knight.

To solve this, you'll need to have the, ? block and turn 4 (the owl), and turn the ?? 5 (the spider).

Once inside, you'll find two chests. The smaller one on the right will have a collection item, while the one straight ahead will have an unidentified head item.

✧ Grand Staircase Tower

Another arithmancy door can be found at the Grand Staircase Tower, just walk up the stairs, and you'll see it.

To solve this puzzle, turn the ?? block directly to 6 (the squid) and the ? block into 7(the lizard).

After you solve the puzzle, you'll find two chests. The one on your right will have a stuffed toy dragon, and the one on the floor will have a Golden Silk Robe.

✧ Grand Staircase

The arithmancy door can be found next to the Grand Staircase fast travel point.

To solve this puzzle, switch the ? block to 0 (the beast) and the ?? block to 5 (the spider).

Once inside, you'll find another two chests. The one on the left will have a collectible item, and the one in the middle will have gear for you.

✧ Library Annex

To get to the arithmancy door, fast travel to the Divination Classroom. Once you do that, walk towards the wooden platform and make a right on the first turn. There'll; you'll find the door.

To solve the puzzle, switch the ? block to the 4 (the owl) and the ?? block to 3 (the snake).

As soon as you get in, you'll find a small chest containing a collectible item on the ground. But wait, there's more! If you follow the hallway down, it'll lead to another room, and inside the room, you'll find a chest with gear and two more arithmancy doors.

✧ Library Annex 2

To solve the arithmancy door near the door in the hidden room, make sure you solve the puzzle mentioned above.

After that, switch the ? block to 4 (the owl) and the ?? block to 5 (the spider).

You'll find two chests inside; one will give you a collectible, and the other will give you gear.

✧ Library Annex 3

You'll need to solve the Library Annex first to get to this door.

Now that you've solved the previous puzzle switch the ? block to 6 (the lizard) and the ?? block to 1 (the unicorn).

When you get inside, you'll find two chests. The small one will give you a collectible item, and the larger chest will give you gear.

✧ Central Hall

The Central Hall arithmancy door can be found near the Potions Classroom fast travel point.

To solve this puzzle, switch the ? block to 8 (the spider) and the ?? block to 3 (the snake).

When you get inside, you'll find two chests. The small one will give you a collectible item, and the larger chest will give you gear.

✧ Potions Classroom

To find this arithmancy door, fast travel to the Potions Classroom. After you do that, walk through the doors and begin making your way downstairs. Once you reach the end, you'll run into a locked level 1 door.

Unlock the door, and inside, you'll find the arithmancy door switch the ? block to 2 (the goat) and the ?? block to 1(the unicorn)

Inside the room, you'll find a Field Guide Page next to the painting, gear in the large chest, and a collectible in the smaller one.

✧ Astronomy Wing

The astronomy wing arithmancy door can be found behind the Charms Classroom fast travel point.

When you arrive at the door, switch the ? block to 2 (the goat) and the ?? block to 7 (the squid).

Once you're inside, you'll find two chests. The smaller one will have a collectible item, and the bigger one will contain gear.

✧ The Great Hall

Fast travel to The Great Hall point. Once you spawn in, make a left towards the windows and walk down the hallway. You'll find the next arithmancy door at the end of the hallway.

To solve the puzzle, switch the ? block to 8 (the spider) and switch the ?? block to 3 (the snake).

Once you're inside, you'll find two chests. The smaller one will have a collectible item, and the bigger one will contain gear.

✧ Hospital Wing

To get to this arithmancy door, you'll have to first play a bit of the main quest. It'll become accessible after completing The Caretakers Lunar Lament mission.

Now that you've gained access to the area, you'll want to walk down to the second floor of the area. Once you've done that, you'll see the door in the corner. To solve the puzzle, switch the ? block 0 (the beast) and the ?? block to 9 (the sea monster).

When you unlock the door, you'll find two chests. The smaller one will have a collectible item, and the bigger one will contain gear.

FIELD GUIDE PAGE LOCATIONS

✧ Hogwarts Castle Field Guide Page Locations

The Astronomy Wing Field Guide Page Locations

There are 23 Field Guide Pages to be found in The Astronomy Wing section of Hogwarts, ranging from simple reveals to more challenging puzzles.

❖ Hogwarts Astronomy Wing Field Guide Page - Wyvern Fountain

■ Requires: Revelio

This fountain in the Transfiguration Courtyard has the likeness of a wyvern on top and serves as a spot for students to gather. You can find it in the center of the Transfiguration Courtyard at the base of the Astronomy Tower, located near the Transfiguration Floo Flame.

❖ Hogwarts Astronomy Wing Field Guide Page - Partial Transfiguration

■ Requires: Revelio

This partially Transfigured teacup serves as a cautionary tale for students who do not take the magic seriously. It can be found on the left side of the Transfiguration classroom, through the door south of the Transfiguration Courtyard Floo Flame.

❖ Hogwarts Astronomy Wing Field Guide Page - Painting of Illyius

■ Requires: Revelio

This painting depicts the young orphan wizard Illyius and his mouse Patronus. You can find it adorning a wall by the stairs at the base of the Defence Against Dark Arts Tower, just Southwest of the Transfiguration Courtyard Floo Flame.

❖ Hogwarts Astronomy Wing Field Guide Page - Serpentine Beast Window

■ Requires: Revelio

A stained glass window depicting a serpentine beast coiled around a tree that extends from the mouth of the legendary Green Man. You can find this window up on the second floor of the Defence Against Dark Arts Tower, over by a skeleton display - and reach it by heading down the stairs from the classroom Floo Flame.

❖ Hogwarts Astronomy Wing Field Guide Page - Augurey Skeleton

■ Requires: Revelio

This skeleton of an Irish phoenix or Augurey can be found in a large display case in the Defence Against Dark Arts Tower. It's located on the same floor as the classroom, over on the west side of the floor past the stairs and the playing instruments.

❖ Hogwarts Astronomy Wing Field Guide Page - Augurey Skeleton

■ Requires: Revelio

This large dragon skeleton that hangs in the Defence Against Dark Arts classroom is allegedly a trophy taken by Professor Hecat after defeating a poaching ring. You can only access it after you've completed your first Defence Against Dark Arts lesson, and you'll find the page on the low balcony.

❖ Hogwarts Astronomy Wing Field Guide Page - Tapestry of Barnabas the Barmy

■ Requires: Revelio

This tapestry depicts Barnabas the Barmy's foolish attempts to teach trolls the art of ballet. You can find near the very top of the Defense Against Dark Arts Tower, up one floor from the Charms Classroom Floo Flame on a short hallway leading to more stairs up to the Astronomy Tower.

❖ Hogwarts Astronomy Wing Field Guide Page - Astronomy Telescope

■ Requires: Revelio

This telescope is the finest stargazing instrument of its kind. It is located at the very top of the Astronomy Wing, above the Defence Against Dark Arts Tower by taking the staircases up past all the classrooms, and even further above the Astronomy Classroom and Floo Flame up to the balconies above.

❖ Hogwarts Astronomy Wing Field Guide Page - Pungent Passage Moth Frame

■ Requires: Lumos

One of the Moth Frames you'll encounter after undertaking the first side quest in the Central Hall. To solve this riddle, you must activate Lumos when facing the painting to reveal where the wayward moth has gone. The frame can be found in the Pungent Passage on the second floor of the Defence Against Dark Arts Tower leading towards the West Tower Floo Flame.

This particular moth is hiding up the hall in the West Tower itself, where there are several tapestries depicting centaurs. Look for the one on the Eastern wall and cast Lumos to summon the moth to you, then take it back down and douse the light to get the Field Page.

❖ Hogwarts Astronomy Wing Field Guide Page - Defence Against Dark Arts Tower Moth Frame

■ Requires: Lumos

One of the Moth Frames you'll encounter after undertaking the first side quest in the Central Hall. To solve this riddle, you must activate Lumos when facing the painting to reveal where the wayward moth has gone. The frame can be found on the fourth floor of the Defence Against Dark Arts Tower, just a bit South of the tower Floo Flame.

This particular moth is hiding one floor up, right outside of Professor Fig's classroom and the Floo Flame here, in front of a large tapestry. Collect it by using Lumos and bring it back downstairs before dousing the light to get another Field Page.

❖ Hogwarts Astronomy Wing Field Guide Page - Astronomy Tower Moth Frame

■ Requires: Lumos

One of the Moth Frames you'll encounter after undertaking the first side quest in the Central Hall. To solve this riddle, you must activate Lumos when facing the painting to reveal where the wayward moth has gone. The frame can be found on the Astronomy Tower's lower landing, located up the stairs from the main tower Floo Flame, below the spiral stairs up to the telescopes.

This particular moth is hiding just a bit higher among all the telescopes on the balcony at the top of the tower Head upstairs to the large moving orbs and look for a chalkboard next to a telescope where the moth hides. Use Lumos to summon it to your wand, then take it back down and turn out the light to get the Field Page.

❖ Hogwarts Astronomy Wing Field Guide Page - Charms Hall Flying Page

■ Requires: Accio

A flying page that you can find outside the Charms Classroom in the Defence Against Dark Arts Tower. This page can be obtained once you have learned Accio from your first day of classes. Outside the Charms classroom, head up the hall to the common area where students are gathered and look for the flying page above them by the stairs up to the Astronomy classroom.

❖ Hogwarts Astronomy Wing Field Guide Page - Dungeons Flying Page

■ Requires: Accio

A flying page that you can technically find under the Bell Tower Wing, but it is listed as an Astronomy Wing page. This page can be obtained once you have learned Accio from your first day of classes. From the Clocktower Courtyard Floo Flame, head south into the North Hall and down the stairs into the dungeons.

Go all the way down the dungeon stairs into the east corridor and turn right to go south all the way into a large dungeon room full of barrels and locked doors. Climb up onto a higher platform on the right, and look for the flying page up above as it loops around stone columns.

❖ Hogwarts Astronomy Wing Field Guide Page - Transfiguration Courtyard Statue

■ Requires: Levioso

A statue holding an orb that you can interact with using the right spell. This page can be obtained once you have learned Levioso from your first day of classes. From the Clocktower Courtyard Floo Flame, head to the East corner under the stone awning to find a statue holding an orb. Use the Levioso spell to raise the orb, and you'll be given a page.

❖ Hogwarts Astronomy Wing Field Guide Page - Astronomy Tower Statue

■ Requires: Levioso

A statue holding an orb that you can interact with using the right spell. This page can be obtained once you have learned Levioso from your first day of classes. Travel to the Defence Against Dark Arts Tower and climb all the way to the Charms Classroom Floo Flame.

From here, head southwest to find the stairs up towards the Astronomy Tower and Classroom, and after the first hallway you will find a statue holding an orb at the base of a spiral staircase. Use Levioso on the orb to raise it, and you can collect the page.

❖ Hogwarts Astronomy Wing Field Guide Page - Astronomy Balcony Dragon Brazier

■ Requires: Incendio

One of the ornate braziers adorned with a small dragon that you can light up around Hogwarts with the proper spell. This page can be obtained once you have learned the Incendio Spell from Professor Hecate. You can find this brazier at the very top of the Astronomy Tower past the classroom on the large balcony where all the telescopes are located.

Look on the right side as you step out onto the balcony for a small staircase leading under the main platform, and here you will find a dragon brazier to light in the corner.

The Bell Tower Wing Field Guide Page Locations

There are 34 Field Guide Pages to be found in The Bell Tower Wing section of Hogwarts, ranging from simple reveals to more challenging puzzles.

❖ Hogwarts Bell Tower Wing Field Guide Page - The Old Librarian

■ Requires: Revelio

One of the first librarians of Hogwarts is depicted in this large portrait. Technically, this field guide is located in the Library Annex. You can find it by entering the Library from the Central Hall to the West, and taking the spiral staircases up to the second floor where the painting is.

❖ Hogwarts Bell Tower Wing Field Guide Page - Three Sisters Bells

■ Requires: Revelio

The Three Sisters Bells are said to be a tribute to the three steep ridges that rise over a glen in Argyllshire. They can be found on display behind a case in the main floor of the Belltower Room, right next to the Floo Flame.

❖ Hogwarts Bell Tower Wing Field Guide Page - Flattened Armour

■ Requires: Revelio

This set of armour belonged to Sir Scagglethorpe the Heedless, and can be found in the main floor of the Belltower Room across from the Floo Flame, flanking the large doors leading out into the courtyard.

❖ Hogwarts Bell Tower Wing Field Guide Page - Broken Broom

■ Requires: Revelio

This broken broom belonged to Selene Wartnaby and is rumored to be all that remains after demonstrating an experimental Lunar Apparition Charm. Like the Flattened Armour, it can be found in the main floor of the Belltower Room across from the Floo Flame, flanking the large doors leading out into the courtyard.

❖ Hogwarts Bell Tower Wing Field Guide Page - Wooden Cat

■ Requires: Revelio

A wooden statue bearing the likeness of Pangur Donn, a fearless feline mouse hunter and devoted study companion. The statue can be found on an upper walkway in the Belltower Hall by a door to the North leading to the Library Annex Hall.

❖ Hogwarts Bell Tower Wing Field Guide Page - Goblin Artefact

■ Requires: Revelio

A horn used by goblins during the 1612 Goblin Rebellion, used to rally troops and annoy witches and wizards. The horn can be found in a display case with two halberds above, on an upper walkway in the Belltower Room located up the stairs South from the Floo Flame.

❖ Hogwarts Bell Tower Wing Field Guide Page - Scorch Marks

■ Requires: Revelio

Allegedly the location of the first known instance of the casting of the Bombarda spell. These markings can be found on a wall in the Southeast upper corner of the Belltower Room up above the Floo Flame, among a row of musical paintings across from the Goblin Artefact.

❖ Hogwarts Bell Tower Wing Field Guide Page - Choir Frogs

- Requires: Revelio

A perch of slimy yet symphonic frogs that comprise the Hogwarts Frog Choir. They are located in a musical room located high up in the Belltower, which you can reach by taking the long stairs in the south corner of the room, or from taking the stairs up in the adjacent North Hall.

❖ Hogwarts Bell Tower Wing Field Guide Page - History of Magic Windows

- Requires: Revelio

This set of stained-glass windows features Merlin, the four Hogwarts founders, and various others of the ages It can be found in the History of Magic classroom, which can be accessed via the North Hall in the Belltower Wing, either traveling South from the Belltower Floo Flame, or West from the Transfiguration courtyard.

❖ Hogwarts Bell Tower Wing Field Guide Page - Urn of Ashes

- Requires: Revelio

This particular urn is rumored to have belonged to a pioneer in dragon taming. It can be found on a display area at the entrance to the Bell Tower Wing's dungeons, which can be accessed by going south from the Bell Tower Courtyard Floo Flame and then down the stairs in the North Hall.

❖ Hogwarts Bell Tower Wing Field Guide Page - Sleeping Dragon Statue

- Requires: Revelio

This particular sleeping dragon statue is made of stone and will never be awoken, despite the Hogwarts motto of never tickling a sleeping dragon. The statue can be found below the Urn of Ashes in the dungeons below the Bell Tower Wing. You can reach it by heading South from the Belltower Courtyard Floo Flame and down the stairs in the North Hall.

❖ Hogwarts Bell Tower Wing Field Guide Page - Werewolf Saga Tapestries

- Requires: Revelio

This set of tapestries tells the tragic tale of a witch bitten by a werewolf. The guide page can be tricky to reach as it is well hidden in the dungeons under the Bell Tower Wing. Start by heading south from the Belltower Courtyard Floo Flame, and down the stairs in the North Hall.

As soon as you enter the dungeons, walk straight into the large tapestry depicting the letter "K" to find a hidden door, and enter the large room beyond that is lined with more tapestries to find the guide page.

❖ Hogwarts Bell Tower Wing Field Guide Page - Bell Tower Flying Page

- Requires: Accio

A flying page that you can find in the main Clock Tower building. This page can be obtained once you have learned Accio from your first day of classes. Return to the Bell Tower Coutyard Floo Flame, and look upwards at the North end of the room above the door to the Library Annex to find a flying page. Grab it with Accio to add it to your collection.

❖ Hogwarts Bell Tower Wing Field Guide Page - Bell Tower Dragon Brazier

- Requires: Incendio

One of the ornate braziers adorned with a small dragon that you can light up around Hogwarts with the proper spell. This page can be obtained once you have learned the Incendio Spell from Professor Hecate. You can find this brazier near the Bell Tower Coutyard Floo Flame. Look in the north corner to the right of the door to the Library Annex to find the brazier, and light it with Incendio to get the page.

❖ Hogwarts Bell Tower Wing Field Guide Page - Castle Ramparts

■ Requires: Revelio

Along with powerful protective enchantments, these defensive ramparts have safeguarded Hogwarts castle for centuries. You can locate this page after visiting Hogsmeade for the first time and being able to explore outside the castle. From the Belltower Courtyard Floo Flame, exit the main doors and look to the right to see the stone ramparts and a small building along the walls you can enter, and look inside for the page.

❖ Hogwarts Bell Tower Wing Field Guide Page - Glumbumbles

■ Requires: Revelio

Glumbumbles are magical flying insects that produce a treacle which will cause melancholy if consumed. Their nests can be found out on the Hogwarts Grounds that you can access after leaving to Hogsmeade for the first time. Exit through the Bell Tower Courtyard and then go East through a small arch towards the outside of the Greenhouses.

Once past the moving dragon topiary, look for a smaller arch past a fountain and turn left to find the hives with the page.

❖ Hogwarts Bell Tower Wing Field Guide Page - Hogwarts Grounds Flying Page

■ Requires: Accio

A flying page that you can find outside the main Clock Tower building. This page can be obtained once you have learned Accio from your first day of classes. Leave the castle via the Clocktower Courtyard large door and head East through the small archway.

As you run past the Greenhouses, look for a flying page that will circle the small fountain here, and use Accio to grab it.

❖ Hogwarts Bell Tower Wing Field Guide Page - Quidditch Pitch

■ Requires: Revelio

The Hogwarts Quidditch Pitch is the site of intense house rivalries, as Chasers, Beaters, Keepers, and Seekers take to the skies in pursuit of the Quidditch Cup.

You can find this Field Guide Page once you are allowed to explore outside of the castle after going to Hogsmeade, and it's a bit of a trek. Exit the Bell Tower Courtyard and go through the large rampart gates West onto the road, and turn left when viewing the Quidditch arena.

Move to the Southwest and stay on the outer side of the ramparts that flank the arena to find a small collapsed part of the wall where the Field Page is hiding.

The Grand Staircase Field Guide Page Locations

There are 26 Field Guide Pages to be found in The Grand Staircase section of Hogwarts, ranging from simple reveals to more challenging puzzles.

❖ Hogwarts Grand Staircase Field Guide Page - Ravenclaw Bust

- Requires: Revelio Spell

This bust created in honour of Ravenclaw house resides in the lofty Ravenclaw Tower. If you chose Ravenclaw as your house, this will be the first Field Guide Page you acquire with Professor Weasley. If not, you can find this by heading West from the Ravenclaw Tower Floo Flame and up the small spiral staircase. Halfway up (before reaching the door to the Common Room), you'll find the bust along the wall.

❖ Hogwarts Grand Staircase Field Guide Page - Ravenclaw Doorknocker

- Requires: Revelio Spell

To gain entrance to the Ravenclaw common room, one must solve a riddle from the door. You can find this by heading West from the Ravenclaw Tower Floo Flame and up the small spiral staircase. Head all the way up to find a large ornate door leading to the tower.

❖ Hogwarts Grand Staircase Field Guide Page - Kelpie Statue

- Requires: Revelio Spell

This statue depicts the Kelpie, a shape-shifting water demon native to Ireland and Great Britain, which usually takes the form of a long-maned horse. If you chose Slytherin as your house, this will be the first Field Guide Page you acquire with Professor Weasely. If not, you can find this at the bottom of the Lower Grand Staircase Floo Flame, heading North down the stairs, and then West back up to reach the statue.

❖ Hogwarts Grand Staircase Field Guide Page - Hufflepuff Barrels

- Requires: Revelio

To enter the Hufflepuff common room, one must tap the barrels in the rhythm of Helga Hufflepuff. If you didn't choose this house, you can find it by going to the Grand Staircase Floo Flame, and turning around to head East down a leafy spiral staircase to the cellar, where you can find the barrels at the very back alcove.

❖ Hogwarts Grand Staircase Field Guide Page - House-Elf Recipe Book

- Requires: Revelio

Rumored to contain some of Helga Hufflepuff's original creations, this book holds a collection of students' favorite recipes. You can find it on the leafy small circular stairwell down to the Hufflepuff Common Room, located just behind the Grand Staircase Floo Flame. It's halfway down the stairwell with an assortment of food you can eat.

❖ Hogwarts Grand Staircase Field Guide Page - The Hogwarts Architect

- Requires: Revelio Spell

A statue of the Hogwarts Architect, surrounded by the four house mascots. It can be found in the Reception Hall between the Entrance and Great Hall that leads towards the Grand Staircase, not far from the Hufflepuff common room stairwell.

❖ Hogwarts Grand Staircase Field Guide Page - Honeydukes Passageway

- Requires: Revelio

Hidden behind the statue of a one-eyed witch is a secret passageway from Hogwarts to the cellar of Honeydukes. You can find it right next to the Faculty Tower Floo Flame that borders the South Wing and

Grand Staircase.

❖ Hogwarts Grand Staircase Field Guide Page - Moving Staircase

■ Requires: Revelio

A savvy student is wise to keep an eye on the stairs, as they will change position without notice. This page can be found along a balcony along the central pillar halfway up the Grand Staircase, past the entrance to Ravenclaw Tower, and before the Arithmancy Door.

❖ Hogwarts Grand Staircase Field Guide Page - Troll Armour

■ Requires: Revelio

This unusual suit of armour was crafted for a troll. It can be found at the entrance to the Trophy Room at the top of the Grand Staircase, located just across from the Trophy Room Floo Flame, and is hard to miss considering its huge size.

❖ Hogwarts Grand Staircase Field Guide Page - Trophy Room

■ Requires: Revelio

This room contains the House Cup, Quidditch trophies, dueling trophies, and other awards given at the school. The page can be found near the center of the main Trophy Room at the top of the Grand Staircase, just past the Floo Flame.

❖ Hogwarts Grand Staircase Field Guide Page - Goblet of Fire Casket

■ Requires: Revelio

The Goblet of Fire rests within this ancient, jewel-encrusted chest. Like the Trophy Room Guide Page, it can also be found within the central Trophy Room just past the Floo Flame at the top of the Grand Staircase, in the center of the room.

❖ Hogwarts Grand Staircase Field Guide Page - Centaur Armour

■ Requires: Revelio

This unique set of armour was created as a misguided peace-offering to a centaur leader. It can be found at the top of the Grand Staircase along the outer edge of the Trophy Room, behind a large glass case.

❖ Hogwarts Grand Staircase Field Guide Page - House-Elf Armour

■ Requires: Revelio

This set of armour for a house-elf is rumoured to have been made by a cruel wizard who wanted his elf to protect him in battle. It can be found at the far edge of the Trophy Room rim at the top of the Grand Staircase over by a locked door.

❖ Hogwarts Grand Staircase Field Guide Page - Ravenclaw Tower Moth Frame

■ Requires: Lumos

One of the Moth Frames you'll encounter after undertaking the first side quest in the Central Hall. To solve this riddle, you must activate Lumos when facing the painting to reveal where the wayward moth has gone. The frame can be found at the base of the Ravenclaw Tower spiral staircase, reached by heading West from the Grand Staircase to the Ravenclaw Tower Floo Flame, and going down.

This particular moth is hiding back up the way you came, in the green room that features an Arithmancy Door near the Floo Flame. Approach the walls and activate Lumos to summon the moth hiding here, and walk back down the stairwell and turn off your light to send the moth to its home and gain a Field Guide Page.

❖ Hogwarts Grand Staircase Field Guide Page - Great Hall Moth Frame

■ Requires: Lumos

One of the Moth Frames you'll encounter after undertaking the first side quest in the Central Hall. To solve this riddle, you must activate Lumos when facing the painting to reveal where the wayward moth has gone. Despite being listed as a Grand Staircase Field Page, The frame can be found in the small antechamber room found between the Reception Hall and Great Hall itself, flanked by various banners and flags.

This particular moth is hiding just in the room ahead, at the far back of the large Great Hall dining area. It's located on the wall behind the owl lectern and staff tables, and you can use Lumos to guide it to your wand. Transport it back to the previous room and turn off the light to get your Field Page.

❖ Hogwarts Grand Staircase Field Guide Page - Grand Staircase Flying Page

■ Requires: Accio

One of the flying pages that you can find floating around Hogwarts. This page can be obtained once you have learned Accio from your first day of classes. Look for the flying page that can usually be found flapping around the central pillar in the staircase, often swooping past the Grand Staircase Floo Flame and then going up a bit higher in its loop.

❖ Hogwarts Grand Staircase Field Guide Page - Quad Courtyard Flying Page

■ Requires: Accio

One of the flying pages that you can find floating around Hogwarts. This page can be obtained once you have learned Accio from your first day of classes. Travel to the Quad Courtyard Floo Flame (which can be reached either by exiting the Great Hall or going through Gryffindor Tower). Here you will find a lonely flying page floating around a large tree in the quad that you can grab.

❖ Hogwarts Grand Staircase Field Guide Page - Ravenclaw Tower Statue

■ Requires: Levioso

A statue holding an orb that you can interact with using the right spell. This page can be obtained once you have learned Levioso from your first day of classes. You can find it at the entrance to the Ravenclaw Tower from the Grand Staircase - or by going East from the Ravenclaw Tower Floo Flame. Use Levioso on the statue with the orb to lift it into the air and claim the page.

Hogwarts - Great Hall Field Guide Page Locations

There are 24 Field Guide Pages to be found in The Great Hall section of Hogwarts, ranging from simple reveals to more challenging puzzles.

❖ Hogwarts Great Hall Field Guide Page - Slytherin Sink

■ Requires: Revelio Spell

Scratched into one of the copper taps on this sink in the girls' toilet is a small snake. You can find this by heading down the stairs from the Lower Grand Staircase Floo Flame, and going through the long passage

East until you come to a set of bathroom doors - and enter the one with the profile of a witch, and inspect the sink in front of you.

❖ Hogwarts Great Hall Field Guide Page - Pear Portrait

■ Requires: Revelio

If one tickles the pear in this still-life painting of a bowl of fruit, it will giggle before turning into a doorknob to allow entry into the Hogwarts Kitchen. If you chose Hufflepuff as your house, this will be the first Field Guide Page you acquire with Professor Weasely. If not, you can find this by going to the Grand Staircase Floo Flame, and turning around to head East down a leafy spiral staircase to the cellar, where you can find the portrait at the cellar entrance.

❖ Hogwarts Great Hall Field Guide Page - House Point Hourglasses

■ Requires: Revelio

These large ornate hourglasses contain rubies, diamonds, emeralds, and sapphires are enchanted to keep the points for Gryffindor, Hufflepuff, Slytherin, and Ravenclaw. It can be found across from the statue of the Hogwarts Architect in the Reception Hall between the entrance and Grand Hall, that leads over to the Grand Staircase.

❖ Hogwarts Great Hall Field Guide Page - Hogwarts Crest

■ Requires: Revelio Spell

The Hogwarts coat of arms includes a lion, snake, eagle, and badger, representing the houses. The large crest can be found as soon you enter the Entrance Hall - either East from the Viaduct Floo Flame outside, or North from the Great Hall Floo Flame.

❖ Hogwarts Great Hall Field Guide Page - The Great Hall Ceiling

■ Requires: Revelio Spell

The ceiling of the Great Hall has been bewitched to mimic the sky above the castle. You can find this guide page right in the middle of the Great Hall, just a few steps from the Floo Flame.

❖ Hogwarts Great Hall Field Guide Page - Owl Lectern

■ Requires: Revelio Spell

This enchanted lectern serves as the spot from which the great headmistresses and headmasters of Hogwarts address the school. It is located at the far end of the Great Hall from its entrance and the Floo Flame spot you can quickly travel to.

❖ Hogwarts Great Hall Field Guide Page - The Yawning Gargoyle

■ Requires: Revelio Spell

Although this smoke-breathing gargoyle might appear to be enchanted, it is actually the Hufflepuff common room chimney. You can find it right outside the doors to the Great Hall dining area to the West, in a small courtyard.

❖ Hogwarts Great Hall Field Guide Page - Slytherin Dungeon Moth Frame

■ Requires: Lumos

One of the Moth Frames you'll encounter after undertaking the first side quest in the Central Hall. To solve this riddle, you must activate Lumos when facing the painting to reveal where the wayward moth has gone. The frame can technically be found off the Grand Staircase past the Slytherin Common Room, over between the two Slytherin bathrooms through the East passage.

This particular moth is hiding a bit further up through the dungeon passages, so head further Southeast and past the spiral staircase to a hall full of locked doors and a large tapestry the moth is hiding next to. Use Lumos to summon it, and take it back to the frame before turning the light off and collecting your Field Page.

❖ Hogwarts Great Hall Field Guide Page - Courtyard Moth Frame

■ Requires: Lumos

One of the Moth Frames you'll encounter after undertaking the first side quest in the Central Hall. To solve this riddle, you must activate Lumos when facing the painting to reveal where the wayward moth has gone. The frame can be found outside the Entrance Hall down a flight of stairs to the North past the courtyard.

This particular moth is hiding back up the way you came, just to the right of the entrance doors to the West by a large Kelpie statue. Use Lumos to attract the moth and walk back down the stairs, and turn the light off to send the moth home and collect your page.

❖ Hogwarts Great Hall Field Guide Page - Great Hall Exterior Flying Page

■ Requires: Accio

A flying page that you can find outside the main Great Hall building. This page can be obtained once you have learned Accio from your first day of classes. Exit the dining area to the large courtyard that the Hufflepuff Common room is built under, and look to the skies to spot a page floating around the middle. Use Accio to nab it for the page.

❖ Hogwarts Great Hall Field Guide Page - Entrance Hall Flying Page

■ Requires: Accio

A flying page that you can find in the Great Hall Wing. This page can be obtained once you have learned Accio from your first day of classes. From the Great Hall Floo Flame, turn and head North past the Reception Hall and into the Entrance Hall, and look up into the rafters of the room to spot a flying page. Use Accio to bring it down, and collect your Field Guide Page.

❖ Hogwarts Great Hall Field Guide Page - Entrance Hall Exterior Statue

■ Requires: Levioso

A statue holding an orb that you can interact with using the right spell. This page can be obtained once you have learned Levioso from your first day of classes. Travel from the Entrance Hall in the Great Hall building to go outside, and then head East under the stone awning to find a statue in the corner. Use Levioso on the orb to raise it, and you can collect the page.

❖ Hogwarts Great Hall Field Guide Page - Entrance Hall Interior Statue

■ Requires: Levioso

A statue holding an orb that you can interact with using the right spell. This page can be obtained once you have learned Levioso from your first day of classes. From the Grand Staircase, enter the Entrance Hall to the left and look along the East Wall for a statue holding an orb. You can use Levioso on it to raise the orb and collect the page.

❖ Hogwarts Great Hall Field Guide Page - Black Lake

■ Requires: Revelio Spell

The Black Lake or Great Lake is an expanse of fresh water south of the castle, and its murky depths are home to merpeople, Grindylows, and other magical creatures. To find this page you'll need to reach the boathouse far below the Great Hall.

Once you've visited Hogsmeade for the first time, the gate to the boathouse will be unlocked along the Northeast part of the Viaduct Courtyard, and you can take the long winding path down the boathouse, where the page will be next to a docked boat.

❖ Hogwarts Great Hall Field Guide Page - Underground Harbour

■ Requires: Revelio Spell

Located deep beneath the Viaduct Courtyard is the landing for boats delivering first-year students across the Black Lake to Hogwarts. In order to find this Field Guide Page, you need to locate a secret lift down to the harbor, which will become active after you've visited Hogsmeade for the first time.

From the Viaduct Courtyard Floo Flame, head North along the lower walkway past a tree to look along the stone walls for a doorway leading to a very small lift elevator. Take it down to the Underground Harbour and inspect one of the boats on the far right of the dock to find the page.

❖ Hogwarts Great Hall Field Guide Page - Great Hall Dragon Brazier

■ Requires: Incendio

One of the ornate braziers adorned with a small dragon that you can light up around Hogwarts with the proper spell. This page can be obtained once you have learned the Incendio Spell from Professor Hecate. You can find this brazier in the Great Hall Dining Room where all the students gather to eat.

In the dining area, look off to the Northwest for stairs up to a second floor storage area, and you'll find a small walkway here with the dragon brazier at the end that you can light up for a field page.

❖ Hogwarts Great Hall Field Guide Page - Boathouse Dragon Brazier

■ Requires: Incendio

One of the ornate braziers adorned with a small dragon that you can light up around Hogwarts with the proper spell. This page can be obtained once you have learned the Incendio Spell from Professor Hecate. To find this brazier you'll need to reach the boathouse far below the Great Hall.

Once you've visited Hogsmeade for the first time, the gate to the boathouse will be unlocked along the Northeast part of the Viaduct Courtyard, and you can take the long winding path down the boathouse, where the brazier will be by the door.

The Library Annex Field Guide Page Locations

There are 20 Field Guide Pages to be found in The Library Annex section of Hogwarts, ranging from simple reveals to more challenging puzzles.

❖ Hogwarts Library Annex Field Guide Page - Portrait of Sir Cadogan

■ Requires: Revelio

Sir Cadogan was allegedly friends with Merlin himself, becoming a Knight of the Round Table, although he can usually be found challenging students to duels. You can find his portrait on the second floor walkway of the Viaduct Entrance to the Library Annex, which can be quickly reached by climbing up the stairs from the Central Hall Floo Flame and then East towards the Viaduct Entrance.

❖ Hogwarts Library Annex Field Guide Page - Palmistry Model

■ Requires: Revelio

This standing model of a hand is demarcated to help students in practice of palmistry. You can find it in the middle of the Diviniation Classroom, which is tucked away at the top of a spiral staircase.

From the Viaduct Entrance of the Library Annex, travel up the spiral staircase on its North side, and head up the staircase past the Floo Flame. A rung ladder will descend as you approach the top, and you can climb it to find the classroom -- and the Field Guide Page.

❖ Hogwarts Library Annex Field Guide Page - Central Hall Fountain

■ Requires: Revelio

This ornate fountain features intricately carved statues of denizens of the magical world. You can find it quite easily after using the Central Hall Floo Flame with Ms. Weasely, and you'll see it adorning the center of the large busy room.

❖ Hogwarts Library Annex Field Guide Page - Statue of Gregory the Smarmy

■ Requires: Revelio

This statue with an ingratiating grin depicts Gregory the Smarmy, who invented the Unctuous Unction potion You can find this statue at the Southwestern entrance to the Central Hall, just to the right of the staircase up to the main fountain - located right next to the Potions Classroom Floo Flame.

❖ Hogwarts Library Annex Field Guide Page - Professor Sharp's Auror Badge

■ Requires: Revelio

This Auror badge belongs to Professor Sharp, and is a symbol of the Ministry's magical law enforcement and protection against Dark Magic. You can find it on a table in the back of the Potions Classroom (easily reached by Floo Flame), off to the side of the Central Hall.

❖ Hogwarts Library Annex Field Guide Page - Greenhouse Tree

■ Requires: Revelio

Situated at the centre of a Hogwarts greenhouse, this giant tree has a large system of roots reaching down to the dungeons. You can find it next to the Greenhouses Floo Flame, which is located off to the Northeast of the Central Hall.

❖ Hogwarts Library Annex Field Guide Page - Dirigible Plums

■ Requires: Revelio

This orange, radish-like fruit floats upside-down as it grows. The tree can be found in the Greenhouses Northeast of the Central Hall, and this tree is located on the ground floor of the easternmost greenhouse through a door.

❖ Hogwarts Library Annex Field Guide Page - Arithmancy Classroom

■ Requires: Revelio

This classroom is where students learn about the magical properties of numbers and numerology. You'll need to solve an Arithmancy Door puzzle yourself to reach it - head up the northern spiral staircase in the Viaduct entrance to the Divination Floo Flame, and exit out onto a wooden walkway.

Follow the walkway to find one of many Arithmancy Doors around Hogwarts - but this one has a chest next to it that holds the cipher. To solve it, reference the number associated with the magical creatures by interacting with the ? and ?? blocks on the wall to equal the circled number in the middle sections.

Once the door is opened, head down the hall into the Arithmancy Classroom to find the Field Guide Page.

❖ Hogwarts Library Annex Field Guide Page - Central Hall Moth Frame

■ Requires: Lumos

The first side quest you can undertake in Hogwarts is also how you learn about getting Field Pages from Moth Frame Paintings. To solve this riddle, you must activate Lumos when facing the painting to reveal where the wayward moth has gone.

This particular moth is hiding off to the Southwest of your current location, by the large statue of Gregory the Smarmy and the large door. With Lumos still active, the moth will follow your wand as you lead it back to the frame, and turn off the light to have it merge with the frame - awarding you with a page.

❖ Hogwarts Library Annex Field Guide Page - Library Moth Frame

■ Requires: Lumos

One of the Moth Frames you'll encounter after undertaking the first side quest in the Central Hall. To solve this riddle, you must activate Lumos when facing the painting to reveal where the wayward moth has gone. The frame can be found on the second floor of the Library on the southern corner.

This particular moth is hiding not far away, at a lectern right in front of the huge portrait of the librarian at the main second floor landing to the Northwest. Use Lumos to guide it to your wand, and bring it back to the frame before dousing your light to claim the Field Guide Page.

❖ Hogwarts Library Annex Field Guide Page - Central Hall Flying Page 1

■ Requires: Accio

One of two flying pages that you can find in the Central Hall of the Library Annex. This page can be obtained once you have learned Accio from your first day of classes. Look for the flying page that spans the length of the large Central Hall, flying high up by the ceiling to swoop down among the higher balconies.

❖ Hogwarts Library Annex Field Guide Page - Central Hall Flying Page 2

■ Requires: Accio

One of two flying pages that you can find in the Central Hall of the Library Annex. This page can be obtained once you have learned Accio from your first day of classes. Look for the flying page that floats around the Southwestern section of the Central Hall, floating around the Statue of Gregory the Smarmy and the large exit door to the Transfiguration Courtyard.

❖ Hogwarts Library Annex Field Guide Page - Viaduct Entrance Flying Page

■ Requires: Accio

One of the flying pages you can find floating around the Library Annex. This page can be obtained once you have learned Accio from your first day of classes. Unlike the two flying around the Central Hall, head back up the stairs from the hall towards the Viaduct Entrance, and climb one of the spiral staircases on the side to spot a page floating high up in the rafters above. It will swoop along the length of the walkway near the Divination Classroom, so be ready to nab it and grab the page.

❖ Hogwarts Library Annex Field Guide Page - Library Flying Page

■ Requires: Accio

A flying page that you can find in the main Library part of the Library Annex. This page can be obtained once you have learned Accio from your first day of classes. Look for the flying page that floats around the second floor of the Library and into the wings on either side before soaring past the middle.

❖ Hogwarts Library Annex Field Guide Page - Central Hall Dragon Brazier

■ Requires: Incendio

One of the ornate braziers adorned with a small dragon that you can light up around Hogwarts with the proper spell. This page can be obtained once you have learned the Incendio Spell from Professor Hecate. You can find this brazier on the far Northern corner of the Central Hall, on an upper balcony overlooking the entrance to the Greenhouses.

❖ Hogwarts Library Annex Field Guide Page - Enchanted Books

■ Requires: Revelio

Best to avoid these books if possible due to their ability to fly from one's hands and contents being of no value. You can find this Field Guide Page on the bottom floor of the Restricted Section of the Library, which you'll unlock during a main quest with Sebastian Sallow. Once you sneak past two ghosts and reach the bottom floor, cast Revelio in the first room to find the page.

❖ Hogwarts Library Annex Field Guide Page - Restricted Section Statue

■ Requires: Levioso

A statue holding an orb that you can interact with using the right spell. This page can be obtained once you have learned Levioso from your first day of classes. However, it is located in the Restricted Section of the Library, which you can only gain access to when undertaking a main quest with Sebastian Sallow. After sneaking down to the lower floors past the ghosts, you'll enter a storage room, and the statue can be found down the stairs. Use Levioso on the statue to lift the orb, and you'll gain a page.

The South Wing Field Guide Page Locations

There are 24 Field Guide Pages to be found in The South Wing section of Hogwarts, ranging from simple reveals to more challenging puzzles.

❖ Hogwarts South Wing Field Guide Page - Portrait of Baruffio

■ Requires: Revelio

This portrait depicts the wizard Baruffio with a large buffalo on his chest. If you chose Gryffindor as your house, this will be the first Field Guide Page you acquire with Ms. Weasley.

If not, you can find this by going to the Faculty Tower Floo Flame in the South Wing, and walking South towards the spiral staircase, and look to the right to find the painting.

❖ Hogwarts South Wing Field Guide Page - Fat Lady Portrait

■ Requires: Revelio

This portrait guards the Gryffindor common room, requiring a password from any who wish to enter. If you did not choose Gryffindor as your house, the easiest way to reach the portrait is by traveling to the Faculty Tower Floo Flame in the South Wing, and walking South to climb up the spiral staircase, and heading down the hall to the West.

❖ Hogwarts South Wing Field Guide Page - Map of Argyllshire

■ Requires: Revelio

This large map depicts Argyllshire, a region in Scotland that is home to black dragons. It can be found on the first floor of the Gryffindor Tower in the South Wing, just across the small bridge North of the Clocktower Courtyard Floo Flame.

❖ Hogwarts South Wing Field Guide Page - Lachlan the Lanky

■ Requires: Revelio

This statue depicts the wizard Lachlan the Lanky - a tall slender wizard who appears quite proud of himself. He can be found at the base of the Gryffindor Tower entrance north across the bridge from the Clocktower Courtyard, and just down the stairs from the Map of Argyllshire in an alcove.

❖ Hogwarts South Wing Field Guide Page - Haunted Toilets

■ Requires: Revelio

A particular bathroom that is one of Peeves's favorite pranking spots. It can be found on the ground floor of the Gryffindor Tower in the South Wing, which you can easily reach by heading North from the Clocktower Courtyard Floo Flame, going down the stairs past the giant map and heading left to find the bathroom hall.

❖ Hogwarts South Wing Field Guide Page - Quad Statue

■ Requires: Levioso

A statue holding an orb that you can interact with using the right spell. This page can be obtained once you have learned Levioso from your first day of classes. Though listed in the South Wing, you can actually find it at the edge of the outdoor Quad area between the Grand Staircase and Great Hall. It's easiest to reach either from the Quad Courtyard Floo Flame and traveling South against the higher walkway. Use Levioso on the statue to raise the orb and collect a page.

❖ Hogwarts South Wing Field Guide Page - Clock Tower Statue

■ Requires: Levioso

A statue holding an orb that you can interact with using the right spell. This page can be obtained once you have learned Levioso from your first day of classes. You can find it right next to the Clock Tower Courtyard Floo Flame, in a small section full of junk off on the right to the main inner courtyard. Use Levioso on the statue to lift the orb

❖ Hogwarts South Wing Field Guide Page - The Well of Four Beasts

■ Requires: Revelio

Some students believe that a wish made over the well of Four Beasts will come true for one who has gained their trust. You can only access after you've traveled to Hogsmeade for the first time and are allowed to leave the school, after which you can leave the Clockwork Courtyard Floo Flame through the large gate and find it outside.

❖ Hogwarts South Wing Field Guide Page - Gryffindor Tower Dragon Brazier

■ Requires: Incendio

One of the ornate braziers adorned with a small dragon that you can light up around Hogwarts with the proper spell. This page can be obtained once you have learned the Incendio Spell from Professor Hecate. You can find this brazier on the first floor of Gryffindor Tower in the South Wing, located in a corner just east of the bathrooms, near the large map of Argyllshire.

❖ Hogwarts South Wing Field Guide Page - Gryffindor Tower Moth Frame

■ Requires: Depulso

Once you have done enough assignments for the Potions teacher to learn the Depulso spell, return to the Gryffindor Tower main floor across the bridge from the Clocktower Courtyard Floo Flame. When inspecting the Map of Argyllshire that had its own Field Guide Page, you can now use Depulso to push in the button above the map, which will reveal a secret room.

Inside, you can find a Moth Frame, and light it up with Lumos to reveal the location of the missing moth.

The painting will reveal the outside of the unlocked Gryffindor Tower bathroom, which happens to be just around the corner from the map of Argyllshire.

You'll have to enter the bathroom, then use Lumos next to the coat rack to lure the moth away. Return to the frame and douse the light to get your Field Guide Page.

✧ Hogsmeade Field Guide Page Locations

Hogsmeade Village Field Guide Pages

There are 55 pages to find all across Hogsmeade Village, including its outskirts and river. These pages usually come in three types:

■ Collection Pages that are invisible until shown from a Revelio spell

■ Flying Pages that require Accio to summon from the air and into your hand

■ Moth Frame Paintings that reveal the location of a lost moth when Lumos illuminates the painting

The pages below are divided up by type, and the order you can usually find them in, as certain buildings become available to explore later in the story.

❖ Hogsmeade Field Guide Page - Hogsmeade

■ Requires: Revelio

The only all-wizarding village in Britain. Hogsmeade has been a favourite haunt of Hogwarts students for centuries. You can find this first Field Guide Page as you approach the Hogsmeade bridge. It's located at the base of a lamp post just to the left of the bridge.

❖ Hogsmeade Field Guide Page - Enchanted Staircase

■ Requires: Revelio

This enchanted staircase in Tomes and Scrolls reveals itself when a particular book is moved. You can find it inside the Tomes and Scrolls shop that is found on the main street as you enter Hogsmeade from the south bridge and look along the left of the street. You can also activate the staircase by touching a book opposite the bookshelf.

❖ Hogsmeade Field Guide Page - Creidwen's Precarious Cauldrons

- Requires: Revelio

The precariously stacked set of cauldrons outside Ceridwen's advertises the shop's wares. It can be easily found as you first enter Hogsmeade from the south bridge entrance, as you'll find the stacked cauldrons and Field Page on the right as you begin to walk down the main street.

❖ Hogsmeade Field Guide Page - Ollivanders Wand Shop

- Requires: Revelio

Wands of a variety of woods and all manner of flexibility, each possessing one of three magical cores, choose their owners in this cosy, cluttered shop run by Gerbold Ollivander. You can find the shop when entering Hogsmeade from the south bridge entrance, heading down the main street, and looking to the left when the main road forks.

❖ Hogsmeade Field Guide Page - Mounted Hog's Head

- Requires: Revelio

This mounted hog's head rolls its eyes and snuffled its nose as it oversees the seedy atmosphere of the Hog's Head pub. You can find this establishment near the West Hogsmeade Floo Flame just down the back alley roads.

Once inside, you'll find the titular head mounted right behind the front counter.

❖ Hogsmeade Field Guide Page - Hog's Head Docks

- Requires: Revelio

These docks outside the Hog's Head Inn offer a pleasant place to sit and relax. You can find it on the other side of the Hog's Head Inn, located along the waterfront by the West Hogsmeade Floo Flame, and the page is hiding near a large cask and crates.

❖ Hogsmeade Field Guide Page - Hogsmeade Community Garden

- Requires: Revelio

This small community garden in the heart of the village is the perfect spot for growing various herbs and magical plants. It can be found just across the small road from the West Hogsmeade Floo Flame, East down some stairs on a small incline behind a row of houses where a garden has sprung up.

❖ Hogsmeade Field Guide Page - J Pippin's Potions

- Requires: Revelio

Established in 1753, J. Pippin's Potions is the apothecary in Hogwarts where the propieter sells a wide variety of potions and ingredients. You can find it in Western Hogsmeade across from the West Hogsmeade Floo Flame, and going inside will reveal the Field Page by the potions desk.

❖ Hogsmeade Field Guide Page - The Three Broomsticks

- ■ Requires: Revelio

Currently run by Sirona Ryan, and as allegedly as old as Hogsmeade itself, the Three Broomsticks is the longstanding gathering spot for villagers and Hogwarts students alike. You can find near the center of Hogsmeade by taking the main road up from the south entrance to where the main road forks. Inside, cast Revelio by the fireplace to find the Field Guide Page.

❖ Hogsmeade Field Guide Page - Butterbeer Barrels

- ■ Requires: Revelio

These barrels contain the popular wizarding drink, Butterbeer. You can find it outside on the North side of the Three Broomsticks building, just across from the entrance to Zonkos, which is near the center of Hogsmeade.

❖ Hogsmeade Field Guide Page - Zonko's Joke Shop

- ■ Requires: Revelio

Frequented regularly by Howarts students and villagers alike, Zonko's is the place to buy pranks and jokes in Hogsmeade. You can find the Field Guide Page at its entrance near the center of Hogsmeade, just to the north of the Three Broomsticks.

❖ Hogsmeade Field Guide Page - Frog Spawn Soap

- ■ Requires: Revelio

An amphibious surprise awaits anyone who tries to wash with this tadpole-filled bar of soap. You can find it inside Zonko's Joke Shop near the center of Hogsmeade, and look for the Field Guide Page by turning left as you enter to look near the front counter for it.

❖ Hogsmeade Field Guide Page - Dungbomb

- ■ Requires: Revelio

A favourite among pranksters, the Dungbomb produces a foul odour, causing anyone in the vicinity to evacuate. You can find this Field Guide Page on the second floor of Zonko's Joke Shop, up near the East door that exits onto the sloped street.

❖ Hogsmeade Field Guide Page - Water Well

- ■ Requires: Revelio

The savvy traveller knows that some water wells may have more to offer than this relatively innocuous water well in the village of Hogsmeade. You can find it along the upper main road near the center of Hogsmeade, just north the Three Broomsticks and Zonko's Joke Shop, in a small courtyard just below Madam Snelling's Emporium.

❖ Hogsmeade Field Guide Page - Gladrags Wizardwear

- ■ Requires: Revelio

Founded in 1750, Gladrags offers a variety of wizarding garments and accessories. You can find this building near the village square by the North Hogsmeade Floo Flame, over on the west side of the square by a large archway. The Field Page can be found out front.

❖ Hogsmeade Field Guide Page - Gladrags Mannequin

■ Requires: Revelio

Unlike the enchanted mannequins favoured by students for dueling practice, the mannequins at Gladrags serve only as a means to display the latest in wizarding fashion. You can find this Field Guide Page inside the Gladrags store by the North Hogsmeade Floo Flame, along the western set of buildings. The mannequin in question is located right by the front door.

❖ Hogsmeade Field Guide Page - Sneakoscope

■ Requires: Revelio

A Sneakoscope is a type of Dark Detector and will spin, whistle, and light up when it detects someone untrustworthy nearby. You can find it in the Dervish and Banges shop, which shares its space with Gladrags on the west edge of the North Hogsmeade Floo Flame and village square. It's located on a table near the entrance door.

❖ Hogsmeade Field Guide Page - Scrivenshaft Cats

■ Requires: Revelio

These cats seem to stay close to Scrivenshaft's quill shop. Perhaps its the feather quills that entice them -- or the owner's treats. You can find this Field Guide Page inside the shop located at the Hogsmeade Central Square by the North Hogsmeade Floo Flame.

❖ Hogsmeade Field Guide Page - Hengist of Woodcroft

■ Requires: Revelio

This statue is of Hufflepuff Hengist of Woodcroft, believed to be the founder of the village of Hogsmeade, after having been driven from his home by Muggle persecutots. You can find this statue in the exact center of the Hogsmeade Square, located by the North Hogsmeade Floo Flame.

❖ Hogsmeade Field Guide Page - Magical Mail

■ Requires: Revelio

❖ Hogsmeade Field Guide Page - Tea Shop Decor

■ Requires: Revelio

Reflective of proprietor Mrs Steepley's taste, this tea shop is awash in the colour pink and frills. This shop stands on the southeastern edge of the North Hogsmeade central square, adjacent to the Honeydukes sweet shop.

The Field Guide Page can be found to the left as you enter the first floor of the tea shop.

❖ Hogsmeade Field Guide Page - Honeydukes

■ Requires: Revelio

Honeydukes sweet shop sells a variety of magical sweets, from Fizzing Whizzbees to Exploding Bonbons and more. This large shop is easy to find, as it's located on the far east side of Hogsmeade next to the North Hogsmeade Floo Flame.

❖ Hogsmeade Field Guide Page - Fizzing Whizzbees

■ Requires: Revelio

These delicious sweets, rumoured to contain Billywig stings, are small sherbert balls that when consumed will levitate one a few inches off the ground. You can find this Field Guide Page as you enter the Honeydukes sweets shop on the east side of Hogsmeade by the North Floo Flame.

❖ Hogsmeade Field Guide Page - Exploding Bonbons

■ Requires: Revelio

These treats explode when eaten. Although unlikely to cause serious injury, one is advised to chew carefully. You can find this Field Guide Page inside the Honeydukes Sweet Shop on the far east side of town by the North Hogsmeade Floo Flame. Once inside, head to the back of the shop where the large Niffler statue is to find the page.

❖ Hogsmeade Field Guide Page - The Magic Neep Cart

■ Requires: Revelio

This cart belongs to Timothy Teasdale, who runs The Magic Neep. You can find this shop on the northwestern outskirts of Hogsmeade, by taking the bridge across from West Hogsmeade Floo Flame to the garden shop, where you'll find the page to the left of the merchant.

❖ Hogsmeade Field Guide Page - Abandoned Shop

■ Requires: Revelio

The proprietor of this abandoned shop is said to have used a Shrinking Charm to aid in storing his inventory. However, a slight miscalculation may have vanished all of it. This shop is located in the far North of Hogsmeade's edge, across the river and Northeast of The Magic Neep at a location called The Old Fool.

❖ Hogsmeade Field Guide Page - The Dogweed and Deathcap Tree

■ Requires: Revelio

This large tree with winding branches supports the shop Dogweed and Deathcap. You can find this shop on the far Northeastern outskirts of Hogsmeade Village, and it's located across the river by taking the bridge up from Hogsmeade Square.

Inside the shop, you can find the Field Guide Page right next to the shop owner and her counter.

❖ Hogsmeade Field Guide Page - Spintwitches Sporting Needs

■ Requires: Revelio

Run by the amiable Albie Weekes, this shop sells all manner of wizarding sporting goods, including the latest broom models - and perhaps even broom enchantments. This shop will only open up after you have had your first flying lessons at Hogwarts by playing through the main story.

After that, you can enter the shop to find the Field Guide Page by a sport uniform.

❖ Hogsmeade Field Guide Page - Flying Page 1

■ Requires: Accio

One of the flying pages found in Hogsmeade that can be summoned to you using the Accio spell. This particular page can be found as you enter the town across the Southern bridge flying to the left in front of the first building.

❖ Hogsmeade Field Guide Page - Flying Page 2

■ Requires: Accio

One of the flying pages found in Hogsmeade that can be summoned to you using the Accio spell. After entering town from the southern bridge, pass the first house on the right and turn right to find a page floating around the backyard.

❖ Hogsmeade Field Guide Page - Flying Page 3

■ Requires: Accio

One of the flying pages found in Hogsmeade that can be summoned to you using the Accio spell. This page can be found on the southwest corner of the region, flying in the grassy field and backyards behind Tomes and Scrolls.

❖ Hogsmeade Field Guide Page - Flying Page 4

■ Requires: Accio

One of the flying pages found in Hogsmeade that can be summoned to you using the Accio spell. You can find it floating around the alley where the Community Garden is just to the east of the West Hogsmeade Floo Flame. It flies back and forth between Ollivanders, Zonkos, and Pippins, and can be grabbed as it floats around the vegetable garden.

❖ Hogsmeade Field Guide Page - Flying Page 5

■ Requires: Accio

One of the flying pages found in Hogsmeade that can be summoned to you using the Accio spell. You can find this one doing loops around the northwestern bridge leading out of Hogsmeade from Pippin's Potions towards the Magic Neep -- just northwest of the West Hogsmeade Floo Flame. Use Accio when it comes out from under the bridge to grab it.

❖ Hogsmeade Field Guide Page - Flying Page 6

■ Requires: Accio

One of the flying pages found in Hogsmeade that can be summoned to you using the Accio spell. You'll find it around the back of the Three Broomsticks near the center of town, in a tiny three-way alley where it swoops in tight circles, and is easily spotted by heading east from Zonko's before turning into the first alley.

❖ Hogsmeade Field Guide Page - Flying Page 7

■ Requires: Accio

One of the flying pages found in Hogsmeade that can be summoned to you using the Accio spell. This one can be found in north central Hogsmeade, located above the Three Broomsticks and above the water well, flying in a small circle around the back of Madam Snelling's Tress Emporium. It can easily be grabbed when looking up from the small water well courtyard.

❖ Hogsmeade Field Guide Page - Flying Page 8

■ Requires: Accio

One of the flying pages found in Hogsmeade that can be summoned to you using the Accio spell. This page can be found a little to the south of the middle northern bridge in Hogsmeade, flying in a low circle around the back of a group of houses - across the water from the cemetery. You can spot it by going up the road north from Zonko's to the bridge and looking left.

❖ Hogsmeade Field Guide Page - Flying Page 9

- ■ Requires: Accio

One of the flying pages found in Hogsmeade that can be summoned to you using the Accio spell. This particular flying page can be found floating around the back of the Brood and Peck Shop near the North Hogsmeade Floo Flame, and northeastern bridge. You can easily spot it by traveling east from the road by the river running past Madam Snelling's Tress Emporium.

❖ Hogsmeade Field Guide Page - Flying Page 10

- ■ Requires: Accio

One of the flying pages found in Hogsmeade that can be summoned to you using the Accio spell. This page can be found making wide loops around the Steepley and Sons tea shop on the southeast corner of the Hogsmeade Square, just south of the North Hogsmeade Floo Flame. Wait for it to swoop around the front of the building to grab it.

❖ Hogsmeade Field Guide Page - Flying Page 11

- ■ Requires: Accio

One of the flying pages found in Hogsmeade that can be summoned to you using the Accio spell. You can find this page soaring around a small park on the far southeast side of Hogsmeade, far south of Steepley and Sons. Once you enter the park with the pink-leaf trees, look for a page circling above.

❖ Hogsmeade Field Guide Page - Flying Page 12

- ■ Requires: Accio

One of the flying pages found in Hogsmeade that can be summoned to you using the Accio spell. This flying page can be spotted soaring around the small pond at the northeastern edge of Hogsmeade, just above the North Hogsmeade Floo Flame. Stand on the edge of the dock and nab the page with Accio.

❖ Hogsmeade Field Guide Page - Flying Page 13

- ■ Requires: Accio

One of the flying pages found in Hogsmeade that can be summoned to you using the Accio spell. This page can be hard to spot, as its flying around one of the farthest buildings on the easternmost side of Hogsmeade. Look on the north side of the house north of Honeydukes for a page to fly around it, and grab it with Accio.

❖ Hogsmeade Field Guide Page - Flying Page 14

- ■ Requires: Accio

One of the flying pages found in Hogsmeade that can be summoned to you using the Accio spell. This flying page is doing loops around the far northeastern bridge in Hogsmeade, on the road leading toward Dogweed and Deathcap across the river. Wait for it to appear from below the bridge and grab it with Accio.

❖ Hogsmeade Field Guide Page - Flying Page 15

■ Requires: Accio

One of the flying pages found in Hogsmeade that can be summoned to you using the Accio spell. You can Field Guide Page circling the Magic Seep, located on the northwestern outskirts of Hogsmeade, by taking the bridge across from West Hogsmeade Floo Flame to the garden shop. Wait for it to circle around the top of the shop before using Accio on it.

❖ Hogsmeade Field Guide Page - Flying Page 16

■ Requires: Accio

One of the flying pages found in Hogsmeade that can be summoned to you using the Accio spell. While technically outside of Hogsmeade, you can find this floating page high up on a ledge to the West of the Magic Neep shop, which is located across the bridge from the West Hogsmeade Floo Flame. Look for some grazing rams near a high ridge to spot it circling around.

❖ Hogsmeade Field Guide Page - Flying Page 17

■ Requires: Accio

One of the flying pages found in Hogsmeade that can be summoned to you using the Accio spell. You can find this page floating around the rocky cliffs along the Northern edge of Hogsmeade, just to the northeast of The Old Fool abandoned building on the outskirts, located between the Magic Neep and the Dogweed and Deathcap shop.

❖ Hogsmeade Field Guide Page - Flying Page 18

■ Requires: Accio

One of the flying pages found in Hogsmeade that can be summoned to you using the Accio spell. This flying page cirlces the Dogweed and Deathcap shop, which is located on the far northeastern edge of Hogsmeade Village across the bridge from the village square. You may want to climb up the ridges that surround the shop for a better shot at grabbing the page as it flies past.

❖ Hogsmeade Field Guide Page - Flying Page 19

■ Requires: Accio

One of the flying pages found in Hogsmeade that can be summoned to you using the Accio spell. This flying page can be found at the far northeastern road leading out of Hogsmeade, aross the bridge and opposite the large water wheel in town. You'll find it circling around the rocks past the river as you pass by the last shop on the outskirts.

South Hogwarts Region Field Guide Page Locations

There are only 3 Field Guide Pages in the South Hogwarts region that encompasses the area around the Hogwarts Castle proper, and they are all based on notable landmarks.

❖ South Hogwarts Grounds Field Guide Page - Groundskeeper's Tools

One might surmise that this set of tools could be enchanted to create all sorts of furnishing for a cosy hut - as large as the occupant may need.

You can find this Field Page outside of Hogwarts just down the road from the Clock Tower Entrance. Cross the long bridge to the South Hogwarts Exit Floo Flame, and look down the road South to spot a small shack in the distance near the water. Head down and enter the house to find the page.

❖ South Hogwarts Grounds Field Guide Page - Spider Parts

It seems someone has been collection Acromantula parts for profit. Lucrative, perhaps, if one is willing to risk one's life to obtain the required inventory.

This Field Guide Page can be found at the small hamlet of Aranshire, just east across the black lake from Hogwarts Castle. In the small village square, look for a stall shopkeeper named Edgar Adley, and inspect the large cage next to his shop to find the Field Guide Page.

❖ South Hogwarts Grounds Field Guide Page - Hogsmeade Station Ticket Office

This station has stood here since the early 1800s when the Minister for Magic acquired a steam engine train to transport students to Hogwarts. The station appeared in Hogsmeade virtually overnight just after the train did.

You can find this Field Guide Page in a small unnamed hamlet north after the bridge between Aranshire and Hogsmeade proper - easily done by following the train tracks. Here at the station you can find a sign for a ticket office on the east side of the platform, and Revelio will show you the Field Guide Page.

Hogsmeade Valley Field Guide Page Locations

There are 4 Field Guide Pages in the Hogsmeade Valley region that encompasses the areas around Hogsmeade itself leading to the Northern bogs and Forbidden Forest.

❖ Hogsmeade Valley Guide Page - Chocolate Frogs

Likely left behind by a Hogwarts student visiting Hogsmeade, these enchanted confections hop around like real frogs. You can find this on a bench over at a park in the far southeast corner of the Hogsmeade Village area.

While technically tracked as a Hogsmeade Valley collectible, it's easier to find this in the Village itself. Take a right after crossing the bridge into Hogsmeade from the South, then turn right and find the park with a balcony that overlooks the valley below to find the page and chocolate left on a bench.

❖ Hogsmeade Valley Guide Page - Pumpkin Fizz

Pumpkin Fizz is a fashionable carbonated drink with a pumpkin flavour. It has yet to become as popular here as Butterbeer, but one never knows.

This Field Guide Page is located in the hamlet of Upper Hogsfield, found North of Hogsmeade Village at the upper edge of the region. Inside the hamlet, travel to the biggest building on the north side that is stocked with barrels of butterbeer and signage, and look for a small front desk in an open room on the left side where the Field Guide Page is hiding.

❖ Hogsmeade Valley Guide Page - Squib Cottage

This idyllic country cottage belongs to an elderly Squib who chose to live in the wizarding world rather than try to integrate with Muggles like some Squibs do.

You can find this large cottage standing along in a field due East of Hogsmeade Village, and just south of the East Hogsmeade Valley Floo Flame. Once spotted from the air, check the south side of the cottage with Revelio to find the Field Guide Page, or head inside to loot some items.

❖ Hogsmeade Valley Guide Page - Runespoor Egg

Produced via the mouth of the three-headed snake, Runespoor eggs are known to enhance mental agility and as such are often used in potions.

This is the only Field Guide Page in the Highlands that requires you to advance to a certain main quest before you can reach it, as it is located deep behind the Falbarton Castle on the far east side of Hogsmeade Valley. You can't access or fly over the castle until you have completed the quest, The High Keep, which is given by Natty during the fall season of the story after your First Trial.

Once you've completed this quest, return to the castle and you'll find that you can now fly over its gates and engage the enemies and Infamous Foe on the far side. If you drop down and look in the ruined building in the furthest northeast corner, you can find the Field Guide Page next to a chest.

North Hogwarts Region Field Guide Page Location

There is only 1 Field Guide Page in the North Hogwarts region that encompasses the area leading from Hogwarts Castle towards the Forbidden Forest.

❖ North Hogwarts Grounds Field Guide Page - Alihotsy Fudge

Made with the leaves of the Alihotsy tree, this fudge is a delightful confection that causes uncontrollable laughter.

This Field Guide Page can be found sitting on a bench overlooking a wide view of Hogwarts from the North, and you can find it on a large bluff overlooking some red-brick ruins you pass by on the road to Hogsmeade. It's located just to the northeast of the Forbidden Forest Floo Flame.

North Ford Bog Field Guide Page Locations

There are only 2 Field Guide Pages in the North Ford Bog region that encompasses the large northernmost area of the Highlands past the Forbidden Forest.

❖ North Ford Bog Field Guide Page - Spider Sign

Evidently a spider infestation has made this area particularly treacherous. One wonders how many areas could use a sign like this.

You'll find this Field Guide Page in the town of Pitt-Upon-Ford, in the far north middle part of the region, just below the large San Barkar's Tower that's hard to miss. Once you land in the hamlet, you'll find the sign (and page) in front the southern bridge that spans the river that runs through the town. It's just south of the merchant Indirra Wolff.

❖ North Ford Bog Field Guide Page - Antique Horn

This oddly-shaped horn was used centuries ago by a Muggle boatman to keep monsters in the water at bay as travellers crossed by boat. Of course, it did no so such thing, as no magical water beast would be repelled by a horn.

This Field Guide Page can be found in the far northeastern reaches of the Bog, on the far right side of the watery bog area that is home to many Duggbog and a few poachers. Look for a small dock Northeast of a Battle Arena and Giant Purple Toad Den to find the horn at the edge of the water.

Hogwarts Valley Field Guide Page Locations

There are 8 Field Guide Pages in the Hogwarts Valley region that encompasses the large central area South of Hogwarts Castle itself, and the river running from the great black lake.

❖ Hogwarts Valley Field Guide Page - Hebridean Black Scale

The Hebridean Black Scale is a large dragon that can grow to over nine metres in length. This large , rough scale belonging to one of the beasts is one of the prized possessions kept in this hideout.

You can find Archie Bickle's tent hideout in a small clearing among the woods due South of Lower Hogsfield, off of one of the main paths going into the valley. Look for a ramshackle looking tent strewn with debris and a large skeleton nearby to find the Field Guide Page.

❖ Hogwarts Valley Field Guide Page - Murtlap Tentacles

Murtlap tentacles are a rare potion ingredient known to raise resistance to certain Dark Arts charms and contain healing properties.

This small container can better be found by looking for a group of standing rocks where some wizards often gather to paint a floating portrait of the nearby Hogwarts Castle across the lake. You can find it along the river's edge by traveling west from the Central Hogwarts Valley Floo Flame, and look on the west side of the river for the clearing among some flat rocks for the tiny vial on the ground.

❖ Hogwarts Valley Field Guide Page - Enchanted Scarecrow

This mischievous-looking scarecrow was long ago enchanted to watch over the garden to ward off crows but now simply harasses gardeners by shouting insults at them while they work.

You can find this Field Guide Page at the hamlet of Brocburrow, located far above the valley on the upper east side of the region, and east of the Central Hogwarts Valley Floo Flame. From the center of town, cross past the Merlin Trial heading west to find a small garden between the two houses on the edge of the hamlet and you'll find the scarecrow there.

❖ Hogwarts Valley Field Guide Page - Ginger Root

The barkeep in Keenbridge keeps a stock of ginger root handy to fend off his customers' nausea - and keep them imbibing a bit longer.

One of three Field Guide Pages found in the hamlet of Keenbridge in lower central Hogwarts Valley, this one can be found just to the right of the Keenbridge Floo Flame, at an open bar counter by the nearby building.

❖ Hogwarts Valley Field Guide Page - Beehives

The wizarding world needs honey for their tea just as much as the Muggle world does.

One of three Field Guide Pages found in the hamlet of Keenbridge in lower central Hogwarts Valley, you can find this one just to the south of the Keenbridge Floo Flame on the other side of the building where the Ginger Roots page is, among a group of large conical beehives along the river.

❖ Hogwarts Valley Field Guide Page - The Tilted House

The locals call this the Tilted House due to the odd angle at which it sits wrapped in the roots of the overhanging tree.

One of the three Field Guide Pages found in the hamlet of Keenbridge in lower central Hogwarts Valley, you can find this page west of the Keenbridge Floo Flame, at the front door to a house with a large tree drooping overhead.

❖ Hogwarts Valley Field Guide Page - Lace Doily

The locals say that this doily was left as tribute to a beloved house-elf who dreamed of one day wearing it

This Field Guide Page lays hidden behind the ruins of an old house due west of the hamlet of Keenbridge up the hill. The site is an Ancient Magic Hotspot, and if you look behind the ruined home, you can find a gravestone at the foot of a tree.

❖ Hogwarts Valley Field Guide Page - Doxy Egg

Stolen by poachers for use in potions, these black eggs come from a Doxy - a magical beast sometimes mistaken for a fairy.

You can find this Field Guide Page inside a small Bandit Camp, which can be found in the hills in the southwest corner of the Hogwarts Valley, between the Keenbridge and Northern South Sea Bog Floo Flames. Follow the paths west and north from the South Sea Bog to find two small bandit camps in the area, and the northern one hides the Field Guide Page in a large open tent.

Feldcroft Region Field Guide Page Locations

There are 7 Field Guide Pages in the Feldcroft region that encompasses the western edge of the Highlands south of Hogwarts, including a large swath of the coastline leading down to the South Bog.

❖ Feldcroft Region Field Guide Page - Peruvian Instant Darkness Powder

When thrown in the air, this powder from Peru creates an impenetrable darkness resistant to most light-creation spells.

You can find this Field Guide page on a bluff high above the West Hogwarts Valley Floo Flame, on the northeastern edge of a Medium Bandit Camp that borders the Feldcroft and Hogwarts Valley regions. Look for a set of tents around an old crumbling building wall to find the darkness powder.

❖ Feldcroft Region Field Guide Page - Broken Binoculars

These broken binoculars were left here by Muggles attempting to track what they suspected was an odd-looking bear but what was likely a Demiguise.

You can find this Field Guide Page on top of some large gatehouse ruins near upper central Feldcroft area, located due north of the Feldcroft village proper. It's located near an Ancient Magic Hotspot, above a large gate where a traveling merchant named Priya Treadwell sells her wares.

❖ Feldcroft Region Field Guide Page - The Feldcroft Well

The well in Feldcroft ran dry some four hundred years ago during a great drought. Sadly, many perished as a result, including at least one young boy whose father refused to speak for years thereafter.

This Field Guide Page is one of two located in the main town of Feldcroft along the western coastline, and it can easily be found in the exact center of the small hamlet where the well is hard to miss.

❖ Feldcroft Region Field Guide Page - Practice Dummies

The residents of Feldcroft sometimes use these to practice their spell-casting -- especially in light of recent goblin attacks.

One of two Field Guide Pages located in this town, the practice dummies can be found just to the east of the town square where the water well lies. Take the road heading out of town to find a line of dummies to the left of a path leading down to a bridge.

❖ Feldcroft Region Field Guide Page - Lovage Bouquet

This memorial serves as a grim reminder of the power of magical beasts. Years ago, a young witch was killed here by a Graphorn as she awaited her true love.

This easy to miss Field Guide Page is located down in a lower alcove along the coastline south of the main Feldcroft Village. Look for a small path leading down to the lower coastal area as you head south, and you'll find a small nook hidden away along the rock walls going southwest towards the water where the shrine sits.

❖ Feldcroft Region Field Guide Page - Cinnamon Bark

It seems that at least one goblin finds the food at this encampment a bit bland, and has taken to adding cinnamon bark to their stew to liven it up.

This Field Guide Page can be tough to locate, as it's actually sheltered under a large rock outcropping that comprises a medium bandit camp that spans platforms both above and below the rock. Fly out southeast from the Feldcroft Catacomb Floo Flame along the coast, then look northward as you curve around the coast to spot the large goblin facility under the rocky cliff.

As you zoom in and engage the goblins under the rock, make your way down to the bottom floor, and you'll find a large brick oven furnace fire area that has the Field Guide Page guarded by more goblins.

❖ Feldcroft Region Field Guide Page - Jewelled Brooch

The gorgeous brooch depicts a magical bird of some sort. Which particular magical bird is unclear, but some believe it's meant to represent one of the medieval Irish druidess Cliodna's birds who sang the sick to sleep.

You can find this Field Guide Page in the hamlet of Irondale, which borders the Hogwarts Valley and South Sea Bog on the lower southeast side of the region. It can be easily found at Irondale's main landmark - the giant windmill that sits on the east side of the hamlet and can't be missed.

South Sea Bog Field Guide Page Location

There is only 1 Field Guide Page in the South Sea Bog region that encompasses the area south of Hogwarts Valley, and is the only passage into the Coastal Cavern leading to the rest of the regions to the southeast.

❖ South Sea Bog Field Guide Page - Abandoned Bothy

This broken-down old bothy used to shelter travelling witches and wizards from the elements. Over time, however, the surrounding bog began to reclaim the bothy.

This Field Guide Page lies hidden in the middle of the large bog area that's found south of the Northern South Sea Bog Floo Flame, and south along the footpath from the bridge where the travelling merchant Priya Treadwell sells her goods. Look for a sinking old shack to find the Field Guide Page outside.

Coastal Cave Field Guide Page Location

There is only 1 Field Guide Page in the Coastal Cave region that encompasses the area dividing the northern and southern regions of the Highlands comprised of a large goblin stronghold that must be traversed to reach the Poidsear Coast and beyond.

❖ Coastal Cave Field Guide Page - Antique Compass

Unfortunately, the fact that this antique Muggle compass was dropped here means the Muggle who dropped it is probably wandering through Wales by now.

Though listed as a Coastal Cavern collectible, you can technically find this Field Guide Page upon successfully making your way through the goblin fortified coastal cave to its southern exit. It's located on a balcony right next to the North Poidsear Coast Floo Flame, overlooking the large region to the south. Since you can't fly through the coastal cave, you'll either have to fight or sneak through the large camp and exit out the tunnel leading south.

Cragcroftshire Field Guide Page Locations

There are only 2 Field Guide Page in the Cragcroftshire region that encompasses the furthest southeastern region of the Highlands, above the Manor Cape and Clagmar Coast.

❖ Cragcroftshire Field Guide Page - Giant Shade Tree

This gorgeous tree has served as home to countless Bowtruckles, fairies, and even perhaps an Augurey or two. Many simply enjoy it for its shade.

You can find this Field Guide page in the only main hamlet of Cragcroft in Cragcroftshire. The main giant tree that stands in the middle of the town located far in the east can't be missed, and the page is at its base.

❖ Cragcroftshire Field Guide Page - Dragon Skeleton

This dragon skeleton may have been here for years - or it could be the result of recent poaching activity. Be wary.

This Field Guide Page can be found along the southern coastline of Cragcroftshire, on a sandy shore to the southwest of the hamlet of Cragcroft, far below the large bluffs, and on the edge of a large bay and bridge that connects to the Clagmar Coast. Look for a large skeleton that's buried among the sand dunes to find the page.

Clagmar Coast Field Guide Page Locations

There are only 2 Field Guide Page in the Clagmar Coast region that encompasses the furthest southern region of the Highlands below both the Manor Cape and Cragcroftshire, thought to be the most dangerous region in these lands.

❖ Clagmar Coast Field Guide Page - Acromantula Venom

Both extremely valuable and extremely poisonous, this venom is secreted from the pincers of the carnivorous Acromantula.

This Field Guide Page is hiding in the corner of a medium-sized bandit camp located on one of the high peaks of the northern part of the Clagmar Coast. Look for the highest bluffs to the left of the bridge and ruins connecting to Cragcroftshire to find the bandit camp, and on its highest platform is a wooden balcony with a table of supplies where the Field Guide Page hides, near a large tent.

❖ Clagmar Coast Field Guide Page - Pungous Onion Bulb

Pungous Onion is a particularly powerful onion used in some potions, including the Cure for Boils. It's best not to touch with bare hands.

This final Field Guide Page is located on a lower path leading down to the southwestern beaches of the Clagmar Coast, not far south from the other Field Guide Page. look for a incline leading down past a small bandit camp on a hill to find a broken stall and scattered supplies on the way to the shoreline, and you'll find the Field Guide Page among the ruins.

COLLECTION CHEST LOCATIONS

✧ Hogwarts Castle Collection Chest Locations

The Astronomy Wing Collection Chest Locations

There are 6 Collection Chests found in The Astronomy Wing section of Hogwarts, ranging from simple locations to Arithmancy Puzzle Doors and Secrets.

❖ Hogwarts Astronomy Wing Collection Chest 1

- Reward: Wand Handle Cosmetic

This chest does not require any special techniques to reach - only knowing where to explore. It can be found in the Defense Against Dark Arts Tower in the Astronomy Wing, found all the way up on the top floor in Professor Fig's Classroom, across the hall from the Floo Flame. You can find it stashed in the left corner of the room opposite the entrance to his office.

❖ Hogwarts Astronomy Wing Collection Chest 2

- Reward: Conjuration Recipe

This chest can be found by solving one of the Arithmancy Doors located all over Hogwarts. Because of this, you'll need the study guide cipher to solve the door puzzle and get to the nearby chest.

While not technically located in the Astronomy Wing, you'll need to look elsewhere to find the entrance. To find it, head down from the Central Hall Floo Flame to the main fountain, and look to the right of the Southern door for an Arithmancy Puzzle Door. Note the ? and ?? blocks on the wall next to it, and on the balcony above. Using the study guide, enter (8) and (3) for the ? and ?? blocks to unlock the door.

Inside will be a random chest as well as a Conjuration recipe chest.

❖ Hogwarts Astronomy Wing Collection Chest 3

- Reward: Conjuration Recipe

This chest can be found by solving one of the Arithmancy Doors located all over Hogwarts. Because of this, you'll need the study guide cipher to solve the door puzzle and get to the nearby chest.

It's located in the Defence Against Dark Arts Tower, on a higher floor where Professor Ronen teaches Charms. In fact, it's just around the corner from the Charms Classroom Floo Flame. Here you'll find a ? and ?? block on either side of the door. To solve the puzzle door, enter (2) and (7) on the ? and ?? blocks to open the door.

Inside you'll find a random chest, and a random Conjuration recipe.

The Bell Tower Wing Collection Chest Locations

There are 6 Collection Chests found in The Bell Tower Wing section of Hogwarts, ranging from simple locations to Arithmancy Puzzle Doors and Secrets.

❖ Hogwarts Belltower Wing Collection Chest 1

- Reward: Conjuration Recipe

This chest does not require any special techniques to reach - only knowing where to explore, and is one of three collection chests located in the upper Belltower area.

Starting at the Belltower Courtyard Floo Flame, look for a small staircase at the south end of the room to climb up the belltower, through narrow stairs and past an Eye Chest.

Once you reach the musical room, look for another small stairway in the far back East side of the room, and continue to climb the belltower past an assortment of bells in the rafters, until you reach a door.

Exit out onto the belltower balcony, and look along the pathway for a small chest that holds the collection item.

❖ Hogwarts Belltower Wing Collection Chest 2 and 3

- Reward: Wand Handle Cosmetic, Conjuration Recipe

These chests do not require any special techniques to reach - only knowing where to explore, and are two of three collection chests located in the upper Belltower area.

Starting at the Belltower Courtyard Floo Flame, look for a small staircase at the south end of the room to climb up the belltower, through narrow stairs and past an Eye Chest.

Once you reach the musical room, look for another small stairway in the far back East side of the room, and continue to climb the belltower past an assortment of bells in the rafters, until you reach a door. Exit out onto the belltower balcony, and look along for a frog statue in one corner.

Interacting with it will transport you to the adjacent belltower balcony. Here you will find not one but two Collection Chests, one of which contains the wand handle, and the other a random conjuration recipe.

The Grand Staircase Collection Chest Locations

There are 7 Collection Chests found in The Grand Staircase section of Hogwarts, ranging from simple locations to Arithmancy Puzzle Doors and Secrets.

❖ Hogwarts Grand Staircase Collection Chest 1

- Reward: Conjuration Recipe

This chest can be found by solving one of the Arithmancy Doors located all over Hogwarts. Because of this, you may want the study guide cipher to solve the door puzzle and get to the nearby chest.

It can be found at the edge of the Reception Hall where it meets the Grand Staircase, on the landing right above the Grand Staircase Floo Flame. However, you'll notice one of the ? blocks are next to the puzzle door while the other ?? block is next to the Floo Flame and the staircase down to the Hufflepuff common room.

Using the study guide, enter (0) and (5) for the ? and ?? blocks to unlock the door. Inside is a small chamber that holds a regular chest, and a Conjuration recipe chest.

❖ Hogwarts Grand Staircase Collection Chest 2

- Reward: Conjuration Recipe

This chest can be found by solving one of the Arithmancy Doors located all over Hogwarts. Because of this, you may want the study guide cipher to solve the door puzzle and get to the nearby chest.

This door can be found at the spiral staircase up to the Ravenclaw Tower common room, which is just Northwest of the Floo Flame for the Ravenclaw Tower. To the right of the door is a ?? block, while the ?

block is hiding back on the opposite wall from the door.

Using the study guide, enter (4) and (5) for the ? and ?? blocks to unlock the door. Inside is a small chamber that holds a regular chest, and a Conjuration recipe chest.

❖ Hogwarts Grand Staircase Collection Chest 3

■ Reward: Conjuration Recipe

This chest can be found by solving one of the Arithmancy Doors located all over Hogwarts. Because of this, you may want the study guide cipher to solve the door puzzle and get to the nearby chest.

This door can be found halfway up the Grand Staircase from the Ravenclaw Tower Floo Flame going up towards the Trophy Room at the top, and you'll find the door along the West wall. The ? block can be found a bit further up the stairs along the wall, while the ?? block is on the central staircase pillar opposite the door.

Using the study guide, enter (6) and (7) for the ? and ?? blocks to unlock the door. Inside is a small chamber that holds a regular chest, and a Conjuration recipe chest.

The Great Hall Collection Chest Locations

There are 3 Collection Chests found in The Great Hall section of Hogwarts, ranging from simple locations to Arithmancy Puzzle Doors and Secrets.

❖ Hogwarts Great Hall Collection Chest 1

■ Reward: Conjuration Recipe

This chest can be found by solving one of the Arithmancy Doors located all over Hogwarts. Because of this, you may want the study guide cipher to solve the door puzzle and get to the nearby chest.

It can be found in the Northeast corner of the Great Hall dining area, down a small hallway on the far side from the Floo Flame. Note the ? and ?? blocks on the walls on either side. Using the study guide, enter (8) and (3) for the ? and ?? blocks to unlock the door.

Inside is a small chamber that holds a regular chest, and a Conjuration recipe chest.

The Library Annex Collection Chest Locations

There are 8 Collection Chests found in The Library Annex section of Hogwarts, ranging from simple locations to Arithmancy Puzzle Doors and Secrets.

❖ Hogwarts Library Annex Collection Chest 1

■ Reward: Conjuration Recipe

This chest is likely the first of many you can find by solving the Arithmancy Doors located all over Hogwarts This is because this is the only door that has the cipher you need to solve the rest located in a nearby chest.

To find it, head to the North side of the Viaduct Entrance in the Library Annex and climb the spiral staircase to the Divination Floo Flame. From here, head out onto a wooden walkway above the Viaduct Entrance, and follow the catwalk to an Arithmancy Door.

Inspect the nearby blue chest to find a cipher that shows each of the animal symbols and their corresponding number. To solve it, you need to select animals for the ? and ?? blocks on the wall that can be added to the other numbers and symbols to equal the circled number in the middle. This particular puzzle requires (4) and

(3). Open the door with the right symbols, and you'll find the collection chest on the other side - plus a path to the Arithmancy Class with more doors.

❖ Hogwarts Library Annex Collection Chest 2

■ Reward: Conjuration Recipe

This chest is likely the one of the first of many you can find by solving the Arithmancy Doors located all over Hogwarts. This is because it's located behind the first Arithmancy Door that has the cipher you need to solve the rest located in a nearby chest.

To find it, head to the North side of the Viaduct Entrance in the Library Annex and climb the spiral staircase to the Divination Floo Flame. From here, head out onto a wooden walkway above the Viaduct Entrance, and follow the catwalk to an Arithmancy Door.

Once you've solved the first Arithmancy Door, head through the passage to the Arithmancy Classroom, where you'll find two more puzzle doors. The one on the left features several symbols with ? blocks on both sides, and you'll need to input (4) and (5) into the ? and ?? blocks to solve it, revealing access to the collection chest plus a chest of random loot.

❖ Hogwarts Library Annex Collection Chest 3

■ Reward: Conjuration Recipe

This chest is likely the one of the first of many you can find by solving the Arithmancy Doors located all over Hogwarts. This is because it's located behind the first Arithmancy Door that has the cipher you need to solve the rest located in a nearby chest.

To find it, head to the North side of the Viaduct Entrance in the Library Annex and climb the spiral staircase to the Divination Floo Flame. From here, head out onto a wooden walkway above the Viaduct Entrance, and follow the catwalk to an Arithmancy Door.

Once you've solved the first Arithmancy Door, head through the passage to the Arithmancy Classroom, where you'll find two more puzzle doors. The one on the far side features several symbols with ? blocks on either side, and you'll need to input (6) and (1) into the ? and ?? blocks to solve it, revealing access to the collection chest plus a chest of random loot.

❖ Hogwarts Library Annex Collection Chest 4

■ Reward: Wand Handle Cosmetic

This chest is one that you can find simply by progressing the main story, as it's hard to miss.

However, it is located in the Restricted Section of the Library, which you can only gain access to when undertaking a main quest with Sebastian Sallow. After sneaking down to the lower floors past the ghosts, you'll enter a storage room, and the chests can be found down the stairs in a far corner of the room.

❖ Hogwarts Library Annex Collection Chest 5 and 6

■ Reward: Wand Handle Cosmetic, Conjuration Recipe

Two chests can be found under the large Viaduct Bridge connecting the Library Annex exterior to the Great Hall and Entrance Hall, but you'll need the Incendio Spell to unlock this Hogwarts Secret.

On the large bridge, you should notice four large braziers (one of which is lit), and each has a large icon in front of it, as well as roman numeral that can be swapped by interacting with the brazier.

On the northwest Library Annex side of the bridge, look along the ground for a puzzle solution. It depicts each of the four symbols found on the braziers, and a roman numeral next to it. This will be your guide. You need to go to each brazier and interact with it until it has the right number, which will only become interactive after you ignite it with the Incendio spell:

- East Brazier - 1

- West Brazier - 2

- South Brazier - 3

- North Brazier - 4

Once all have been correctly numbered and ignited, the puzzle on the floor will open to reveal a ladder down below the bridge. Here you'll find an assortment of chests, which include a Wand Cosmetic, Conjuration Recipe, and Legendary Chest.

The South Wing Collection Chest Locations

There are 5 Collection Chests found in The South Wing section of Hogwarts, ranging from simple locations to Arithmancy Puzzle Doors and Secrets.

❖ Hogwarts South Wing Collection Chest 1

- Reward: Wand Handle Cosmetic

This chest does not require any special techniques to reach - only knowing where to look and what to interact with. Enter the base of the Gryffindor Tower, which you can do by going North across the small bridge from the Clockwork Courtyard Floo Flame.

Once in the building, head right up the stairs and past the assortment of musical portraits until you reach a spiral staircase. There is a frog statue at the base of the stairwell that you can interact with, which will instantly teleport you to a different room.

Look for a chest on a table in this room to gain a new cosmetic Wand Handle, and exit the far door to fall back down into the hallways with musical portraits.

Hogsmeade

There are 5 Collection Chests to find in Hogsmeade.

❖ The Three Broomsticks

- Location: The Three Broomsticks

From the South Hogsmeade Floo Flames, head northwest up the road to The Three Broomsticks. Go inside and go up the stairs on the left. Follow the stairs as they wind up to a locked door (Requires Alohomora to unlock). Inside, the Collection Chest is against the wall on the right.

North Hogwarts Region

There are 7 Collection Chests to find in Hogwarts Grounds.

❖ Bandit Camp #1

- Location: Small Bandit Camp

You'll visit this Bandit Camp during the course of the Main Quest The Helm of Urtkot. After dealing with the enemies, the Collection Chest can be found inside the blue tent.

❖ The Collector's Cave

■ Location: The Collector's Cave

You will enter The Collector's Cave during the Main Quest The Helm of Urtkot. After defeating the first group of enemies (Inferi), continue through the tomb. After making your way through the third moth door, continue to the left. In the next room, look right and cast Depulso on the doors to blast them open. Next, use Accio or Wingardium Leviosa to move the crate over to the right. Climb up to find the Collection Chest.

South Hogwarts Region

There are 9 Collection Chests to find in South Hogwarts Region.

❖ Lower Hogsfield #1

■ Location: Lower Hogsfield

From the Lower Hogsfield Floo Flames, head into the village and run along the edge of the lake. You'll find this Collection Chest under an upturned cart.

❖ Lower Hogsfield #2

■ Location: Lower Hogsfield

From the Lower Hogsfield Floo Flames, head into the village. Follow the cobbled path to the left and around to the back of the house. This Collection Chest is sat on some crates near the Vendor.

✧ Hogsmeade Village Collection Chest Locations

Hogsmeade Collection Chests

There are 5 Collection Chests to be found in total in the village of Hogsmeade, but not all are available right away.

❖ Hogsmeade Collection Chest 1

■ Reward: Conjuration Recipe

One of the first Collection Chests you can find in Hogsmeade, but you'll still need to progress the story past The First Trial, until you have learned the Alohomora Spell from Mr. Moon when the season changes to fall.

Once you have gained the level 1 unlocking spell, return to The Three Broomsticks and climb to the very top of the tavern. There's a level 1 locked door leading to a Private Room, and inside you'll find the Collection Chest, along with a Demiguise Statue and Field Guide Page.

❖ Hogsmeade Collection Chest 2

■ Reward: Conjuration Recipe

This Collection Chest will require you to have the maximum rank of Alohomora, which will require you to turn in 9, and then 13 more Demiguise Moons in order to get Rank 3 of the unlocking spell.

Once you have the highest Alohomora Spell rank, travel to the South Hogsmeade Floo Flame, and look around the side of the house it's placed at to find a level 3 door lock. Inside, you'll find the Collection Chest stashed away among other treasure chests.

❖ Hogsmeade Collection Chest 3

■ Reward: Conjuration Recipe

This Collection Chest will require you to have the maximum rank of Alohomora, which will require you to turn in 9, and then 13 more Demiguise Moons in order to get Rank 3 of the unlocking spell.

Once you have the highest Alohomora Spell rank, travel along the southeast back roads of Hogsmeade to find the shop called Flutes and Lutes. Walk along the past just to the southwest of the Flutes and Lutes shop to find a smaller home with a level 3 door lock, and use Alohomora to enter and find another Collection Chest with some other treasure chests.

❖ Hogsmeade Collection Chest 4

■ Reward: Conjuration Recipe

This Collection Chest will require you to have the maximum rank of Alohomora, which will require you to turn in 9, and then 13 more Demiguise Moons in order to get Rank 3 of the unlocking spell.

Once you have the highest Alohomora Spell rank, go to the Central Hogsmeade Floo Flame located at the village square, and travel east past Honeydukes to a group of two houses along the ridge. The northernmost of these two houses has a level 3 lock on the door, and you can use Alohomora to gain entry. Inside you'll find the Collection Chest.

❖ Hogsmeade Collection Chest 5

While this Collection Chest does not require any unlocking spells, you will need to advance the story until you find a quest that has you tracking an Ashwinder Hideout located in a cellar below the Hog's Head Inn.

ALL HOGWARTS SECRET PUZZLE SOLUTIONS

Hidden throughout Hogwarts Legacy are several secrets that you can uncover to receive valuable loot, such as legendary gear, wand cosmetics, Field Guide Pages, and even Conjuration recipes. Now, there's a good chance you have already stumbled across many of these secrets; however, the process needed to complete them is often quite complex and will involve multiple steps, special spells, side quest completions, and much more.

This guide will explain everything you need to know about solving all of Hogwarts' biggest secrets that are available to uncover.

✧ How to Find and Solve All Hogwarts Secrets

How to Solve the Viaduct Bridge Puzzle

❖ Hogwarts Library Annex Collection Chest 5 and 6

■ Reward: Wand Handle Cosmetic, Conjuration Recipe

Two chests can be found under the large Viaduct Bridge connecting the Library Annex exterior to the Great Hall and Entrance Hall, but you'll need the Incendio Spell to unlock this Hogwarts Secret.

On the large bridge, you should notice four large braziers (one of which is lit), and each has a large icon in front of it, as well as roman numeral that can be swapped by interacting with the brazier.

On the northwest Library Annex side of the bridge, look along the ground for a puzzle solution. It depicts each of the four symbols found on the braziers, and a roman numeral next to it. This will be your guide. You need to go to each brazier and interact with it until it has the right number, which will only become interactive after you ignite it with the Incendio spell:

- East Brazier - 1

- West Brazier - 2

- South Brazier - 3

- North Brazier - 4

Once all have been correctly numbered and ignited, the puzzle on the floor will open to reveal a ladder down below the bridge. Here you'll find an assortment of chests, which include a Wand Cosmetic, Conjuration Recipe, and Legendary Chest.

How to Solve The Key of Admittance Secret Puzzle

❖ **The Key of Admittance - Headmasters Office Secret**

■ Reward: Field Guide Page, Wand Handle, and Conjuration Recipe

After the completion of the Niamh Fitzgeralds Trial main quest, you'll gain to the Headmaster's Office, which is located above the Trophy Room.

From the Headmaster's Office, look for the door that's beside the desk with a telescope next to it - on this desk, you'll find a chest that contains a Wand Handle.

Using the Alohomora spell, unlock the level 2 lock and continue up the set of stairs. At the top, you'll encounter another locked door, this time level 3.

When opening the door, it'll reveal a room with a key on the desk - pick this up.

When entering the room, you'll also unlock the Room with a View Trophy.

Using the newly collected Key of Admittance, return to the Headmaster's Office and make your way down the spiral staircase until you reach the hallway.

Follow the hallway to the end, where you'll come face-to-face with a large metal door with a keyhole in the middle.

When opening the door, you'll find a large wooden loot chest and a spiral staircase that will lead you to uncover two Collection Chests and a Field Guide Page.

HOW-TO GUIDES

✧ Floo Flames Guide: How to Fast Travel

With so much to do and explore, it's easy to become lost and overwhelmed by the sheer size of the map in Hogwarts Legacy. So, you'll be happy to know that fast traveling is an option, and it's available as soon as you reach Hogwarts. This guide will explain everything you need to know about unlocking and using the fast-traveling network known as Floo Flames in Hogwarts of Legacy.

How to Use Floo Flames to Fast Travel

Once you arrive at Hogwarts, you'll discover Floo Flames during the opening mission. These fast travel points – indicated by a bright green flame on the wall – can be found all across Hogwarts and when exploring the grounds outside the campus known as The Highlands.

To use the Floo Flames, you'll first need to ensure that you have them unlocked. To unlock a Floo Flame, you'll need to visit their location. Thankfully, you don't need to interact with Floo Flames to unlock them, as simply nearing their surrounding area will have them becoming discovering.

An unlocked Floo Flame is represented by a green flame on the map, whereas an inactive Floo Flame will appear as grey.

Once you've unlocked Floo Flames, simply open your map and select any of the green Floo Flame icons you wish to travel to, and then select the fast-travel option to be teleported to its location. With little load times, fast traveling has never been easier in the wizarding world.

✧ How to Change Your Gear Appearance (Transmog)

Hogwarts Legacy is chock full of gear, whether it be neckwear, cloaks and robes, outfits, and even facewear. But if you're looking to keep appearances, mixing and matching gear to ensure you're receiving the best stats possible won't always provide the best appearance. Thankfully, Hogwarts Legacy utilizes a transmog system, which allows you to essentially activate cosmetic overrides for all types of gear.

So, if your wizard fancies themself a fashionista, this guide will explain everything you need to know about using the transmog system to change your gear appearance in Hogwarts Legacy.

How to Use Transmogs to Change Gear Appearance

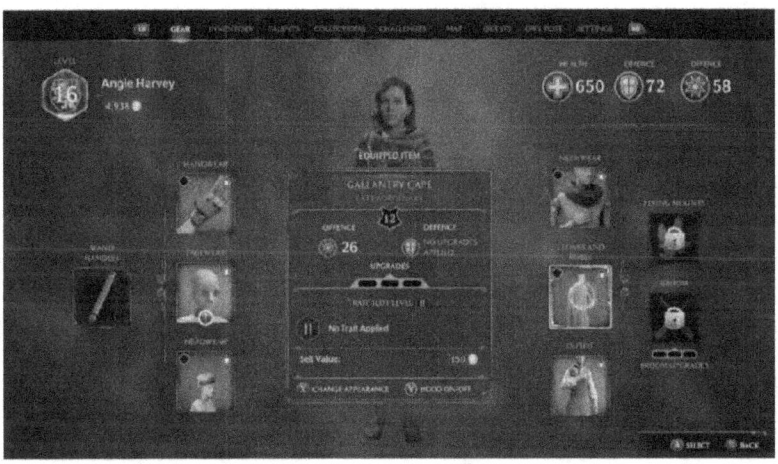

To change your gear appearance, you will first need to equip an item. Once equipped, hover over the item slot and select the "Change Appearance" option. This will now bring up the Appearance Menu, which will allow you to assign a different visual to pieces of gear you own.

In this menu, you'll find two different sections: Collections and Other. Collections are unique appearances that you have unlocked through the completion of challenges and will appear within the Appearances section of the Change Appearance menu – there is a total of 88 appearances to unlock through the completion of challenges and quests.

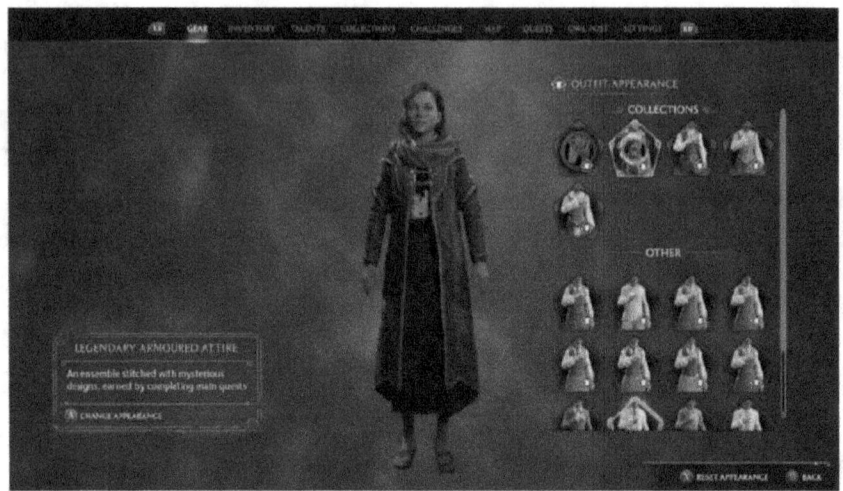

The Other section is as described, as these items will be pieces of gear you have found when searching chests.

Gear transmogs are purely cosmetic and will not affect your character's gear stats.

Once you have changed the appearance of your gear, return to the Gear menu, and you'll notice that an eye symbol has appeared over the gear slot – this indicates that a transmog is currently active. Do note, though, that each time you change a piece of gear, you will also need to adjust the transmog override.

✧ How to Open Arithmancy Puzzle Doors

If you've explored Hogwarts, there's a good chance you have encountered the large wooden number doors that are surrounded by strange beast symbols and locked behind mathematical equations. These puzzle doors are known as Arithmancy Doors, and while they may seem intimidating at first glance, they're actually relatively easy to solve once you learn what the symbols represent. This guide will explain everything you need to know about solving the Arithmancy Door puzzles in Hogwarts Legacy.

How to Find the Arithmancy Number Door Cipher

Please note that you do not need to uncover the Symbol Cipher to solve the Arithmancy Door puzzles. So skip ahead to our solution if you don't wish to find the cipher note.

To find the Arithmancy Door cipher, start by traveling to Divination Classroom Floo Flame, which is located in The Library Annex of Hogwarts.

From this Floo Flame, head through the doorway to your right and travel across the wooden walkway until you reach an intersection – turn right and follow the path to the end.

At the end of the path, search the small blue chest on the bench seat next to the blackboard. Here, you'll discover the Arithmancy Door cipher, which reveals the numerical value of each symbol represented around the door's archway.

Now, solve the Arithmancy Door nearby, and it'll lead you to the Arithmancy Classroom, where students learn about the magical properties of numbers and numerology. In the classroom, you'll also find two additional Arithmancy Puzzle Doors that both lead to treasures.

How to Solve the Number Puzzle Doors

As we explained earlier, finding the cipher isn't necessary to solve the Arithmancy Door Puzzles. As long as you know the numerical value of each symbol, you can unlock all Arithmancy Doors you encounter.

So, whether you've obtained the cipher or not, solving the Arithmancy Door puzzles can seem a little intimidating, but the solutions are much easier than they look. To solve the puzzle, simply approach the door and reveal a set of equations. Both equations will feature a central number that is surrounded by three figures.

All three outer numbers must equal the central number when added together.

Here's the catch, though, some of the symbols will be replaced by a question mark and potentially a creature-like symbol, which will match one of the symbols that can be seen around the arch of the door. Each of these symbols represents a number, and starting from left to right, the symbols will represent the numbers 0 through 9.

Fortunately, with a simple substitution equation, you'll have the answer needed to unlock the door.

Using the above door as an example, you'll always want to start from the central number. So, we'll start with nine, and then you'll want to subtract both two and the dragon-like symbol, which, when using the cipher, shows that its numerical value is three.

● 9 - 2 - 3 (Dragon) = ?

This means that the ? is equal to four – which is represented by the bird symbol.

● ? = 4 (Bird)

To confirm this result, add all three of these numbers together, and you'll get nine.

Now, simply repeat this step for the second equation – see the equation solution below:

- 4 - 1 - 0 = ??

  - ?? = 3 (Dragon)

Now that you've worked out the value of ? and ??, roll the large triangle symbols beside each door, so the question marks represent the symbol of each numerical value.

Once the symbols are rolled into place, open the door, and you'll now have access to all the hidden loot inside – which will mostly consist of gear.

✧ How to Increase Your Gear Inventory Space

Inventory management will play a significant part in your Hogwarts Legacy experience, as right from the start, you're left to work with just 20 gear slots. While it may seem like a lot, you'll soon learn that loot chests are everywhere, and with gear stats increasing as you level up and progress through the game, you'll find yourself constantly destroying and selling off gear pieces. This guide will explain everything you need to know about increasing your gear capacity in Hogwarts Legacy.

How to Increase Your Gear Inventory Capacity

To tackle this issue, you'll want to unlock additional gear inventory slots by completing Merlin Trials, which are unlocked after completing the mandatory main quest, Trials of Merlin. This quest is quite a few hours into the game, so you'll need to do a little bit of inventory management until you reach this stage.

During this mission, you'll learn everything you need to know about solving the trials, which are a series of unique puzzles that require the use of different spells, and at times, the surrounding environment in order to solve them.

Once you have completed the quest, you'll notice that Merlin Trials will have populated all over your world map.

Each completed trial will count towards the "Complete Merlin Trials" Exploration Challenge. Every milestone, which starts at two trial completions, will unlock four additional inventory slots. Should you complete all the milestones, you'll end up with a total of 40 available gear slots.

To unlock the initial milestone, we recommend completing the Merlin Trials closest to Hogwarts and Lower Hogsfield, as they're some of the easier trials to complete in terms of requirements and the overall difficulty.

It's important to note, that you must claim the rewards for each completed milestone, as the additional inventory slots will not be automatically added.

✧ How to Change Your Character Appearance

There's nothing worse than creating a character and not being able to adjust their style later down the track. Thankfully, Hogwarts Legacy allows you to change not only your hairstyle and hair color but also your complexion, eye color, and much more. So, if you're looking to spice things up or simply want to adjust your style, here's everything you need to know about changing your appearance in Hogwarts Legacy.

How to Change Your Character Appearance

To change your appearance in Hogwarts Legacy, make your way to Madam Snelling's Tress Emporium, which is found in the northwestern corner of Hogsmeade.

When entering the hairdressing salon, speak with Calliope Snelling and ask to view her services. Here, you'll be presented with a menu similar to the character creation system you initially used to create your character.

While you won't be able to adjust your face shape, you will have access to change up the following features:

- Hair Color

- Hairstyle

- Complexion

- Freckles and Moles

- Scars and Markings

- Eye Color

- Eyebrow Color

- Eyebrow Shape

Once you have made your changes, select confirm -- Madam Snelling's service will set you back 20 Galleons -- and you'll be on your way with a snazzy new do.

✧ How to Unlock Eye Chests

Eye Chests are non-standard chests you'll encounter many times in Hogwarts Legacy. While opening them early in the game seems impossible, they're pretty simple once you've acquired a particular basic yet helpful spell. This page contains information on how to unlock Eye Chests and what mission you need to complete to start opening them.

How to Unlock Eye Chests

After arriving in Hogwarts, you'll soon run into one of these sentient chests with an eye in the middle. They can be found at every corner of the game—from the Restricted Section of the Library in Hogwarts to inside shops in Hogsmeade.

Approaching the chests will startle them, causing the eye to follow your every move, indicating that you'll need to sneak up on them to open them. However, your standard crouch won't be enough, but when combined with the Disillusionment Charm, it will prove effective in unlocking the chests.

How to Get The Disillusionment Charm

While most basic spells are acquired by attending classes, you'll learn the Disillusionment Charm from Sebastian Sallow during the main quest, Secrets of the Restricted Section. He'll teach you the charm by the Central Hall to sneak past some students and into the Library.

When you make it to the Restricted Section, you will be able to use the charm to open an Eye Chest for the first time. Before you get too close to the chest, you want to cast the spell and get close enough until you are prompted to unlock it.

- Note that if you alert the Eye Chest and try to go invisible, the chest won't return to its dormant state—you'll need to walk away, cast the charm, and sneak up to open it.

Rewards from Eye Chests

Though most chests in and out of Hogwarts contain randomized gear, collections, and other items, Eye Chests will always have 500 Gold inside—making them an easy source of quick cash.

The majority of them can be found either in Hogwarts or in Hogsmeade. Be sure to cast Revelio whenever you're out exploring; you never know where these chests might pop up.

✧ Alohamora Guide: How to Open Locks

If you've been exploring the many points of interest in Hogwarts Legacy, you've probably encountered locks on doors highlighted in blue by Revelio. These locked doors can lead to valuable equipment, collectibles, and secret passageways you might not be able to find elsewhere. Luckily, there is a specific charm you can learn to help you pick these locks and open up previously inaccessible places in the game.

How to Open Locks in Hogwarts Legacy

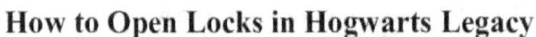

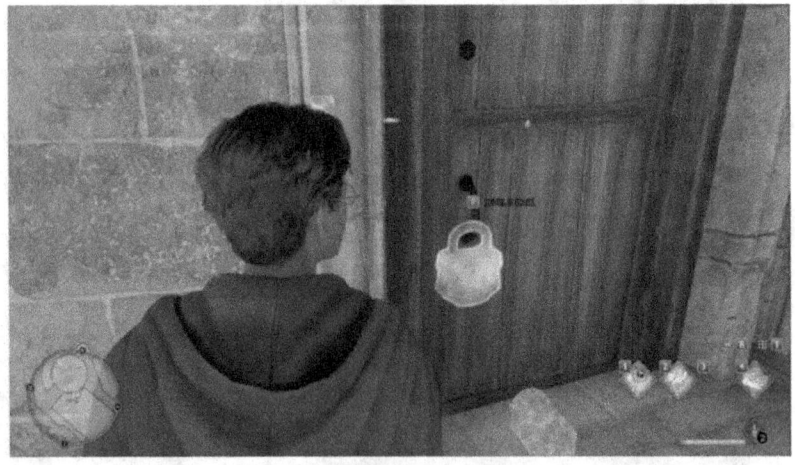

Like many spells and upgrades in the game, you'll need to progress far enough in the main questline to acquire them; the same goes for the charm required to pick these locks—the Alohamora Charm.

Also known as the Unlocking Charm, Alohamora is the last Essential Spell you can unlock in Hogwarts Legacy. There are two main quests you'll need to complete to learn the charm:

1. Percival Rackham's Trial: After finding out the location of the Map Chamber underneath Hogwarts,

you'll be prompted to complete the first of Professor Rackham's four ancient magic trials. Completing this quest will open the second mandatory quest to unlock Alohamora.

2. The Caretaker's Lunar Lament: Once you're back in Hogwarts, make your way to the Reception Hall and talk to Gladwin Moon to start the quest. The quest starter is just above the Grand Staircase Floo Flame location. Mr. Moon will teach you the Alohamora Charm to get into a restricted castle section during the quest.

Acquiring all 9 Demiguise Statues will also complete the second quest from Mr. Moon, The Man Behind The Moons. As a reward, Mr. Moon will teach you both advanced levels of Alohamora that'll allow you to pick Level II and Level III locks—which often contain far better loot than those of Level I locks.

How to Use Alohamora

To use Alohamora, approach and interact with a lock to cast the Unlocking Charm—this will trigger a lockpicking mini-game you need to crack to open the door.

The mini-game is relatively straightforward; all you need to do is move the inner and outer circles slowly until you get to the point where the gears in the bottom right (Green) and the middle (Red) start rotating.

■ Keep the circles in place for a second, and the lock should be picked, allowing you to get inside the door and grab whatever is inside.

- You can also use Alohamora on locked chests.

✧ How to Get and Upgrade Your Broom

What better way to explore a world filled with magic than to fly on an enchanted broom? To make traveling in Hogwarts Legacy less time-consuming, you will need your own broom. Nevertheless, you must complete several prerequisites before you can fly in and out of the castle whenever you'd like.

This guide will cover all the necessary information on how to get a broom, including what missions to complete, where to purchase one, and how to upgrade it.

How to Get and Use a Broom

To get a broom, you'll need to progress in the main story until you've completed the quest Jackdaw's Rest where you find the Tap to Reveal underneath Hogwarts.

Once you've completed that quest, you'll be prompted to attend your first Flying Class by the main lawn. Finish the two tasks Madam Kagawa gave you, and then go on a lap around the castle with Everett Clopton.

After finishing the class, fast travel to any Hogsmeade and make your way to the Spintwitches Sporting Needs shop across from Ollivanders.

■   You'll notice that the shop is closed if you have tried to go before attending your first Flying Class.

Speak to Mr. Weekes inside and browse his selections of available brooms—each costing you 600 Gold. If you don't have enough gold, you can sell some unwanted gear or go to Gladrags Wizardwear and open the Eye Chest using the Disillusionment Charm for an easy 500 Gold.

To use your broom, ensure you are not inside a No-Fly Zone, as indicated by the symbol to the right of your minimap.

Equip your broom by holding LB/L1/Tab (Xbox, PS, PC) and pressing B/○/3. Moving while flying has the same controls as your basic movement, though you can use a temporary speed boost by pressing LT/L2/Left Mouse Click.

How to Upgrade Your Broom

Buying your first broom will also unlock the side quest, Flight Test, from Mr. Weekes—he'll advise you to meet Iselda Reyes at the Quidditch Pitch.

- Completing Iselda's time trials will reward you with upgrades to your broom, improving your overall flying experience.

✧ How to Get Mandrake and Venomous Tentacula

Trying to complete Professor Garlick's Assignment 1 side quest in order to unlock the Wingardium Leviosa spell in Hogwart Legacy? Well, there's a good chance you're probably wondering how you can find the two requested plants: Venomous Tentacula and Mandrake.

Thankfully, despite no assistance from your world map, locating the Dogweed and Deathcap is relatively easy. So, here's everything you need to know about finding a Mandrake and Venomous Tentacula for Professor Garlick.

How to Find Dogweed and Deathcap

So, in order to get Mandrake and Venomous Tentacula, you'll need to visit the Dogweed and Deathcap store, which you'll find along the northern outskirts of Hogsmeade, directly across the river from Madam Snelling's Tress Emporium.

The Dogweed and Deathcap store is represented by a mushroom icon on the map.

How to Get Mandrake and Venomous Tentacular

Once you arrive at the Dogweed and Deathcap store, head inside and speak with Madame Beatrice Green to view her stock. Here are the following plants, seeds, and tools that she sells:

- Seed Packets
  - Chinese Chomping Cabbage Seed - 600 Galleons
  - Mandrake Seed - 800 Galleons
  - Venomous Tentacula Seed - 1,050 Galleons
- Room of Requirement Items
  - Fertiliser (2) - 300 Galleons
- Combat Tools
  - Chinese Chomping Cabbage (2) - 300 Galleons
  - Mandrake (2) - 500 Galleons
  - Venomous Tentacular (2) - 600 Galleons

✧ How to Solve the Bell Tower Puzzle

How to Solve the Bell Tower Puzzle in Solved by the Bell

After starting the Solved by the Bell side quest, you can find the Bell Tower Puzzle above the Choir

Room in the Bell Tower Wing, on the western half of Hogwarts; the closest Floo Flame location to the room would be the Bell Tower Courtyard.

From the courtyard, go up the first set of stairs to your right until you get to a tiny corridor with an Eye Chest at the end—the next room over will be the Choir Room. Go to the back of the room and up the wooden stairs, and you'll make it to the Bell Tower.

While you can try your hardest to solve it using Accio, having the Wingardium Leviosa Charm will make putting back the two missing bells a breeze—as you can fine-tune the height, distance, and rotation of the object you're levitating.

To learn the spell, you must progress further in the main questline until you attend your first Herbology Class with Professor Garlick. She will then give you the assignment to test the Venomous Tentacula and the Mandrake plants in combat.

Complete the assignment and return to her class; she will teach you how to use the advanced levitation charm there.

Check out our guide on how to obtain those two plants here.

Return to the Bell Tower and use the charm on the two bells. The first bell will be right on the set of stairs, while the second is tucked in a corner below the second flight of stairs. If you're struggling to locate them, use Revelio, and it'll highlight them in blue.

The first bell must go in the row below, next to the biggest one. The second will fit in the row above, all the way to the left.

Since you're already in the Bell Tower, check out what's at the top. On the balcony, you'll find a small chest plus a Toad Statue that'll teleport you to another tower's balcony containing a legendary chest.

✧ How to Change Spells

How to Change Spells

In Hogwarts Legacy, you can only have four active spells on your action bar; you can see which are currently active on your bottom right screen.

To change them, press the Right D-Pad (Consoles) or T (PC) to access all your unlocked spells. From there, hold RT/R2 (Xbox & PS) and use the D-Pads to choose which set to edit, then hover over the spell or charm you want to equip and press the corresponding slot button to activate it unto your action bar.

For Mouse and Keyboard users, simply drag and drop the spells in whichever slot you like.

During combat, you can access the different Spell Sets by scrolling up or down (PC); or holding RT and pressing the corresponding D-pad on the controller.

How to Increase Spell Sets

Your spells are your best friends in Hogwarts Legacy—they are the difference makers during combat, the solutions to many puzzles, and will make your overall experience in the game that much easier. But having only four active spells at a time can get dull quickly.

Fortunately, you'll be able to add two extra sets of spells directly at Level 5. However, you must complete the main quest Jackdaw's Rest, to unlock the Talent tab.

If you've completed the quest at a later level, the talent points you've gained up until that point will all be automatically available.

Once you've completed Jackdaw's Rest, open your Talent tab and go to the Core talent tree. You'll want to spend your talent points for Spell Knowledge I and Spell Knowledge II under the Level 5 talents.

You'll be able to get the last Spell Set by unlocking Spell Knowledge III at Level 16.

In total, you can equip 16 total spells at a given time. Try to utilize the sets for different purposes, i.e., one for your main combo, one for utility, one for the Room of Requirement, etc.

✦ How Leveling Up Works in Hogwarts Legacy

What Leveling Up Does in Hogwarts Legacy

As you gain experience from undertaking certain tasks in Hogwarts Legacy, your character will begin to level up from 1 to 40. Though it not may not be immediately apparent how leveling helps your character - it's important to remember you're still just a 5th year student at a school.

As you level up, you'll gain the following bonuses:

- Maximum Health Gain (the amount gained each level will increase the higher your level gets).

- Talent Points (The Talent Tree will be unlocked after completing the Main Quest, Jackdaw's Rest, and will retroactively add talent points for all previously gained levels.

- The ability to equip higher level gear (and find new gear at or around your current level.

- Unlocking of certain side quests and activities.

As you can see, leveling is important to becoming a more powerful character in Hogwarts Legacy, but the main quest recommended levels are often low enough that you'll be okay going for awhile without leveling up in certain areas, as you'll still gain spells regardless.

Gaining Experience - The Field Guide Book

Unlike most RPGs, simply running around and defeating monsters or dark wizards won't always give you the experience you're looking for to grow in power. In fact, you won't level up at all during the Prologue of the game even after getting into some nasty fights.

It's not until you're sorted into one of the four Hogwarts Houses and Professor Weasley gives you a quick tour that you'll be introduced to the conduit for your learning experience - the Field Guide Book. It can be pretty easy too overlook the brief tool-tip that spells it all out: Experience is gained by completing Field Guide Book Challenges. What may not be as clear is that these Challenges are the ONLY way to gain experience in Hogwarts Legacy. This is because all the major activities you take part in - from main questlines to collectible hunting - feed into Field Guide Book Challenges, buy only to a certain point.

This means that as you defeat enemies, grab collectibles, or complete quests that relate to a certain challenge in your Field Guide Book, you'll gain experience for each enemy or item that is tallied towards the challenge goal - as well as meeting the goal itself.

However: Once you have fully completed that challenge, you will not get ANY more experience for continuing, which includes defeating more of that enemy type, finding more of that collectible in the region, or whatever that challenge required. For example, while there are 150 Field Guide Pages in Hogwarts Castle alone, you'll only gain experience for collecting them for the challenge tiers that add up to 100 pages total.

Any pages gained after that will not grant any experience, but you can still gain more by collecting Field Guide Pages in Hogsmeade or The Highlands regions - to a certain point.

This will prevent you from simply grinding on one type of experience gain to overlevel past a certain point. This also means that if you want to level up as high and as fast as you can, you'll need to vary your activities to include all available options as they appear through the story.

List of Experience Gaining Challenges

The list below includes all the types of challenges that you can gain experience from undertaking - though many will be unlocked as you progress further into the story, though others can be undertaken during any point in your adventures:

Combat Challenges	Quest Challenges	Exploration Challenges	Field Guide Page Challenges	Room of Requirement Challenges
Defeat Dark Wizards	Complete Teacher Assignments	Collect Ancient Magic Traces	Collect Field Guide Pages in Hogsmeade	Breed Unique Beasts
Defeat Dugbogs	Complete Main Quests	Pop Balloons	Collect Field Guide Pages in Hogwarts	Rescue Beasts
Defeat Goblins	Complete Side Quests / Relationship Quests	Locate Landing Platforms	Collect Field Guides Pages in The Highlands	Upgrade Your Gear
Defeat Inferi		Complete Merlin Trials		
Defeat Infamous Foes		Find Astronomy Tables		
Defeat Spiders		Solve Hogwarts Secrets		
Defeat Trolls				
Defeat Mongrels				

✧ How to Upgrade Gear

How to Upgrade Gear

In the early portion of the game, upgrading your gear is as simple as swapping your old equipment for new and improved ones. But as you progress in the main questline, you'll unlock the Room of Requirement to use as your haven to craft potions, grow plants, and, most importantly, upgrade gear.

After completing the main quest, The Helm of Urtkot near Hogsmeade, go back to the Room of Requirement, and Deek will show you how to use the Nab-Sack to rescue Beasts from poachers during The Elf, the Nab-Sack, and the Loom quest.

Completing this quest will unlock two main conjurations in the room for upgrading your gear.

The first is the Vivarium, a magical green-house that acts as a free-range for all the Beasts you've rescued In the Vivarium, you can interact with Beasts by feeding and brushing them, allowing you to collect valuable materials to upgrade different gear parts.

- Owning the Vivarium is key to upgrading your gear, as purchasing individual materials from Broof and Peck in Hogsmeade will cause hundreds of gold.

The second is the Enchanted Loom, the tool you'll use to upgrade your gear directly. To conjure an Enchanted Loom, use your Conjuring Spell, go under the Utility tab, and then put it wherever you please.

Gear Upgrades: Stats vs. Traits

Each piece of gear features one primary Offence or Defence stat by default. You can add and upgrade a secondary stat or apply Traits to your gear by using the magical materials you gather from the beasts with the Loom.

Traits are bonuses you can apply to your gear, but you must unlock them through challenges or acquire them from Bandit Camps. In addition, you can only put Traits on Superb rarity gear and above.

- For example, Extraordinary equipment will have a Level II Trait slot, while a Legendary will have a Level III (Max) one.

The first three species you've collected should suffice for upgrading the first level. However, you'll need rarer beasts for advanced upgrades.

While exploring the surrounding Highlands, look for the pawprint symbol on your map for beast lairs—it'll indicate which species of beast you can find there. Be sure to use our Region Guides to find specific locations of beast lairs in all the regions.

✧ How to Get a Hippogriff

How to Get a Hippogriff

Before you can get a Hippogriff, you'll need to progress far enough in the main questline until you get to attend your first Beast Class.

You'll first encounter a Hippogriff after that class. This Hippogriff, named Highwing, was befriended by Poppy Sweeting after she rescued the beast from poachers.

Soon after that class, speak to Deek in the Room of Requirement to obtain the Nap-Sack, a magical item that allows you to rescue and store beasts in it. After learning how to use the Nap-Sack, keep following the main questline until you get to the quest, The High Keep.

Follow the quest objectives to infiltrate Falbarton Castle with Natsai Onai. Halfway into the quest, you both find Highwing captured by Theophilus Harlow's men.

Rescuing Highwing

Unfortunately, due to its well-sought feathers, Hippogriffs are incredibly vulnerable to poachers, and Highwing was captured once more alongside its paired half.

Continue the quest's objectives; climb up the castle all the way up, and be wary of the Dark Wizards patrolling the vicinity—they are most likely the most formidable enemies you've encountered thus far.

■　For a more in-depth guide on the quest, be sure to check out our walkthrough here.

On one of the castle floors, you can rescue two Puffskeins and a Niffler trapped inside a cage room. Since they're all in a small enclosure, saving them using a Nap-Sack won't be difficult.

Once you get to the top of the castle, a cutscene will play out where you and Natsai free the two Hippogriffs. From there, you'll hop on Highwing to escape Harlow's men and fly above the Hogsmeade Valley.

Once you finish the quest, you can summon her whenever you roam the open world by holding LB/L2/Tab (Xbox, PS, PC) and pressing Y/△/2.

✧ How to Get Fluxweed Seeds

How to Get Fluxweed Seeds and Fluxweed Stems

To find Fluxweed Seeds or Fluxweed Stems, you'll want to visit The Magic Neep. This store can be found across the river in the northwestern corner of Hogsmeade. Here, you will find Timothy Teasdale, Hogsmeade's supplier of fresh produce, seeds, and fertilizer.

The Magic Neep is essentially your one-stop shop to unlock all types of plants, seeds, and ingredients needed to grow plants and craft potions for the Herbology class or the Room of Requirement.

❖ The Magic Neep - Shop List

Here are all the seeds and ingredients that you can purchase from The Magic Neep:

Seed Packets

Fluxweed Seed – 350g

Knotgrass Seed – 350g

Mallowsweet Seed – 200g

Shrivelfig Seed – 450g

❖ Room of Requirement Items

Fertiliser – 300g

❖ Ingredients

Dittany Leaves – 100g

Fluxweed Stem – 150g

Knotgrass Sprig – 150g

Mallowsweet Leaves – 100g

Shrivelfig Fruit – 150g

How to Get Focus Potions - J Pippin's Potions

Now, once you've picked up the Fluxweed, you're probably going to want to start crafting your own potions in the Room of Requirement. To do so, though, you'll first need to pick up some potion recipes from J. Pippin's Potions - this store can be found in the northwestern corner of the Hogsmeade, just before traveling over the bridge that leads to The Magic Neep.

It is here that you can purchase the following recipes, ingredients, and combat tools needed to craft potions:

❖ Potion Recipes

Focus Potion Recipe - 1200g

Thunderbrew Recipe - 1200g

Invisibility Potion Recipe - 800g

Maxima Potion Recipe - 500g

❖ Ingredients

Ashwinder Eggs - 150

Dugbog Tongue - 100g

Horklump Juice - 50g

Lacewing Flies - 100g

Leaping Toadstool Caps - 150g

Leech Juice - 150g

Spider Fang - 50g

Stench of the Dead - 100g

Troll Bogeys - 100g

Mongrel Fur - 50g

❖ Combat Tools

Focus Potion - 500g

Thunderbrew - 1000g

Edurus Potion - 300g

Invisibility Potion - 500g

Maxima Potion - 300g

Wiggenweld Potion - 100g

✧ How to Get Troll Bogeys

Where to Buy Troll Bogeys

If you've got some Galleons to spare, Troll Bogeys can be purchased from J. Pippin's Potions in northwest Hogsmeade. Found under the Ingredients section of the store, these rather disgusting ingredients will set you back 100 Galleons each.

How to Find Troll Bogeys

However, if you're strapped for Galleons, your next best option will be to hunt down Trolls and collect Troll Bogeys that way. Trolls can be found marked on the map with a cave-like icon in regions such as Feldcroft, Coast Cavern, Marunweem Lake, Glagmar Coast, and more.

If you're looking for recommendations, we suggest searching for Trolls in northern Fledcroft or the Coastal Cavern regions, as they're relatively easy to reach.

Once you have collected the Troll Bogeys, return to Professor Onai in order to unlock Decendo.

See our handy Professor Onai's Assignment guide for a complete step-by-step walkthrough.

✧ How to Open the Main Gate

How to Open the Main Gate to Falbarton Castle

After your conversation with Natty is over, head forward and to the right to climb up the battlements. Hit the mechanism on the left with Depulso, then Wingardium Leviosa the crate inside over to the right and then Levioso it, climb on top and climb again to reach the upper rampart.

Hit the wooden boards to free a crawlspace, but before you crawl through, head around the right and use Accio through the window on the crate in the room. Return to the crawlspace and head through.

Once inside, hit the gate mechanism multiple times with Depulso, and then quickly use Accio to pull on the ring on the wall to block the gate from fully closing.

With Natty joining you, walk with her up to the gate to witness a cutscene where Highwing the hippogriff breaks out as dark wizards hit it with spells.

✧ How to Get a Large Pot

How to Purchase a Large, Medium, and, Small Pot

In order to purchase plant pots, make your way to Tomes and Scrolls, which is found in the southern area of Hogsmeade Village.

When talking to the shopkeeper, you'll have the opportunity to purchase plant pots, along with a number of Conjuration recipes for the Room of Requirement, which will allow you to create potions, herbology tables, and more.

❖ Tomes and Scrolls - Available Items

Tomes and Scrolls will stock and sell the following Conjuration recipes:

■ Beast Feeder Spellcraft:

1200g

- Beast Toybox Spellcraft:

  500g

- Chopping Station Spellcraft:

  1500g

- Dung Composter Spellcraft:

  100g

- Hopping Pot Spellcraft:

  3000g

- Material Refiner Spellcraft:

  1500g

- Potting Table with One Large Pot Spellcraft:

  1000g

- Potting Table with Two Large Pots Spellcraft:

  3000g

- Potting Table with One Medium Pot Spellcraft:

  750g

- Potting Table with Two Medium Pots Spellcraft:

  1500g

- Potting Table with 3 Medium Pots Spellcraft:

  3000g

- Potting Table with Three Small Pots Spellcraft:

  400g

- Potting Table with Five Small Pots Spellcraft:

  2500g

- Medium Potions Station Spellcraft:

  1000g

- T-Shaped Potions Station Spellcraft:

  2000g

✧ How to Advance Time

To advance time, simply view the world map, and you'll find a small "Wait" option that can be toggled by pressing your controller's analog stick, or the respective button on PC. Selecting this option will allow you to change the game from day to night.

Stock in many stores is limited, but did you know that if you advance time by two and a half days, their stock will be fully replenished. This is a great method if you're looking to purchase bulk seeds, ingredients, or potions.

✧ How to Find Diricawl

How to Find Diricawl

Since Professor Howin's Assignment will be before you explore the southern portion of the Highlands, you probably don't have access to most of the available Diricawl Dens in the game unless you've done a lot of off-course exploration.

Thankfully, there is one Diricawl Den that is a short distance away from Hogwarts. From Hogwarts Castle, head south towards the Keenbridge hamlet in Hogwarts Valley, the area directly south of the South Hogwarts Region.

The Diricawl Den is on the cliff northwest of the hamlet, next to an Ancient Magic Hotspot, a Merlin Trial, and an entrance to a Treasure Vault.

How to Catch a Diricawl

Once you've arrived at the Diricawl Den, get rid of the several Inferius spawning nearby before trying to catch a Diricawl—it'll make doing the task much more manageable.

Equip the Spell Set with your Nab-Sack in it, along with either of these two spells:

- Glacius (Madam Kogawa's Assignment 1)

- Arresto Momentum (Madam Kogawa's Assignment 2)

The first two spells are the key here—as previously mentioned, the Diricawl can teleport short distances, making Arresto Momento and Glacius the perfect spells to keep them in place.

For Professor Howin's Assignment, it doesn't matter whether you catch a male or female Diricawl, though it might be easier to go for a female one since their vibrant color doesn't blend with the surrounding.

There is usually four Diricawl in this particular den, consisting of two males and two females. Cast Glacius or Arresto Momentum as soon as you are within range of one, as they'll immediately teleport away when you get too close.

- Once you freeze them in place, run towards them and use your Nab-Sack as fast as possible.

The Diricawl has four meters you need to fill with the Nab-Sack; the two spells mentioned above should give you enough time to keep them in place before they attempt to teleport again.

ALL PLANTS AND INGREDIENTS

Combat is a huge part of Hogwarts Legacy. After all, danger lurks on every corner. In this game, you can use plants during combat to keep your enemies at bay, but there's more; certain plants will also be essential ingredients for brewing potions or crafting gear.

This guide has all the information you need about plants in Hogwarts Legacy. Including their names, what they do, and how to use them, growing requirements, and much more.

✧ How Plants Work in Hogwarts Legacy

Plants have many uses, but in Hogwarts Legacy, you will mainly use them to brew potions, craft better gear, or during combat encounters.

You can grow your plants inside the Hogwarts castle, either during the Herbology class or in the Room of Requirement. Each plant will have a timer that uses real-time to mark how many minutes you will need to wait before the plant is ready to harvest.

✧ All Plants and Seeds

Below is a complete list of all the plants and seeds that are available to purchase, as well as their cost price, growing requirements, and what they are used for. All ingredients can be grown at the Potting Station in the Room of Requirement.

All Plant and Seeds					
Plant/Herbology (Combat)	Use	How to Find	Cost Price	Growth Time	Plant Size Requirement
Chinese Chomping Cabbage	Will follow and attack enemies until destroyed.	Herbology class; you can also purchase seeds at Dogweed & Deathcap and grow them yourself.			Grown in medium or large pots.
Venomous Tentacula	When deployed, they will shoot venom at nearby enemies.	You can purchase seeds at Dogweed & Deathcap and grow them yourself.			Grown in large pots.
Mandrake	Mandrakes are sentient plants with a root that can cry, this sound is fatal to anyone who hears it, but a Mandrakes seedling's cry only causes unconsciousness.	You can purchase seeds at Dogweed & Deathcap and grow them yourself.			Grown in pots of any size.

✧ All Potion Ingredients

Below is a complete list of all the potion ingredients that are available to find and purchase, as well as their use, where you can find them, and how much they cost.

All Potion Ingredients

Potion Ingredients	Used to Create	How to Find	Cost Price
Lacewing Flies	Focus Potion	These can be found around the world or purchased at The Magic Neep or J. Pippins Potions in Hogsmeade.	
Fluxweed Stem	Focus Potion	These can be found around the world or purchased at The Magic Neep or J. Pippins Potions in Hogsmeade.	
Dugbog Tongue	Focus Potion	Kill Dugbogs around the school or purchase at The Magic Neep or J. Pippins Potions in Hogsmeade.	
Leech Juice	Thunderbrew Potion Maxima Potion	These can be found around the world or purchased at The Magic Neep or J. Pippins Potions in Hogsmeade.	
Shrivelfig Fruit	Thunderbrew Potion	These can be found around the world or purchased at The Magic Neep or J. Pippins Potions in Hogsmeade.	
Stench of the Dead	Thunderbrew Potion	These can be found around the world or purchased at The Magic Neep or J. Pippins Potions in Hogsmeade.	
Ashwinder Eggs	Edurus Potion	Kill Ashwinder's outside of the school or purchase at The Magic Neep or J. Pippins Potions in Hogsmeade.	
Mongrel fur	Edurus Potion	Kill Mongrel's outside of the school or purchase at The Magic Neep or J. Pippins Potions in Hogsmeade.	
Spider Fang	Maxima Potion	Kill Spider's outside of the school or purchase at The Magic Neep or J. Pippins Potions in Hogsmeade.	
Horklump Juice	Wiggenweld Potion	These can be found around the world or purchased at The Magic Neep or J. Pippins Potions in Hogsmeade.	
Dittany Leaves	Wiggenweld Potion	These can be found around the world or purchased at The Magic Neep or J. Pippins Potions in Hogsmeade.	

| Mandrakes

Mandrakes are sentient plants with a root that can cry, this sound is fatal to anyone who hears it, but a Mandrakes seedling's cry only causes unconsciousness. You can use the Mandrake to incapacitate enemies.

Dittany

Dittany is a plant used in potion-making due to its powerful healing and restorative powers.

In Hogwarts Legacy, you'll be able to grow and collect Dittany.

Horklump Mushroom

These mushrooms can be collected to use in Wiggenwald Potion.

Chinese Chomping Cabbage

One of the plants you will be able to use during combat is a red cabbage-like plant that you launch onto your enemies, dealing damage.

Shrivelfigs

In one of the trailers, we can see one of the optional companions, Sebastian Swallow, trying to use what looks like a Shrivelfig to try and reverse a curse.

Venomous Tentacula

You can use the leaves of this plant in potion making, but if you ever find yourself in trouble, you can use Venomous Tentacula's spiked vines to attack your enemies.

Mallowsweet

Mallowseet is a magical herb that Centaurs burn to inhale its fume, which helps them to stargaze.